Airworthiness and HUMS Specialists' Meeting 2018

Huntsville, Alabama, USA
21 – 22 February 2018

ISBN: 978-1-5108-7001-7

Printed from e-media with permission by:

Curran Associates, Inc.
57 Morehouse Lane
Red Hook, NY 12571

Some format issues inherent in the e-media version may also appear in this print version.

Printed by Curran Associates, Inc. (2018)

For permission requests, please contact Vertical Flight Society
at the address below.

Vertical Flight Society
2701 Prosperity Ave, Suite 210
Fairfax, VA 22031
USA

Phone: (703) 684-6777
Fax: (703) 739-9279

www.vtol.org

Additional copies of this publication are available from:

Curran Associates, Inc.
57 Morehouse Lane
Red Hook, NY 12571 USA
Phone: 845-758-0400
Fax: 845-758-2633
Email: curran@proceedings.com
Web: www.proceedings.com

TABLE OF CONTENTS

Flexible Integrated Intelligent Network for Prognostic Health Management Systems

W Thomas Little
Electrical Engineer
AVNIK Defense Solutions, Inc.
Huntsville, AL, USA

Michele Kochoff Platt
CEO/President
AVNIK Defense Solutions, Inc.
Huntsville, AL, USA

Jeff Yalowitz
CEO
Instrumental Sciences, Inc.
Huntsville, AL, USA

ABSTRACT

Weapons systems depend on the health of their components to perform reliably over extended periods of time. Maintenance is an important aspect of reliability, assuring equipment is available when needed. Trends toward electronic/networked information present opportunities to improve maintainer effectiveness, but new tools must be developed to manage interactions between the information sources and maintainers. AVNIK, with subcontract team member ISI, researched techniques for Flexible Integrated Intelligent Network (FIIN) for Prognostic Health Management (PHM) systems, based on the artificial intelligence field of cooperative multi-agent distributed work environments. AVNIK also developed prognostics algorithms using statistical anomaly detection and trend analysis. Our approach applies a multi-agent architecture to produce a distributed information management toolset.framework enabling the sharing of system health information to aircrew members, aircraft to transmit inflight aircraft health/diagnostics information to ground crews for maintenance operations and allow for the prepositioning of assets and maintenance execution. These tools gather relevant information from diverse sources upon user request or autonomous notification, filter information, and format it for presentation to users.

INTRODUCTION

The US Army Aviation and Missile Research Development Center (AMRDEC) Aviation Development Directorate (ADD) funded AVNIK and subcontract team member, ISI to develop and conduct a feasibility and trade-off assessment of a FIIN, under Phase I SBIR Topic A16-090, "Flexible Integrated Intelligent Network (FIIN) for Prognostic Health Management (PHM) Systems". This paper includes initial results from AVNIK developing and demonstrating the FIIN system, an autonomous intelligent network architecture, using new technologies of multiple cooperative intelligent agents (CIAs) and actual Apache aircraft fault data.

Our system design included CIAs, which are portable/mobile software programs that encapsulate Artificial Intelligence (AI) techniques, such as the ability to incorporate knowledge bases and decision-making engines. These AI techniques permit the CIAs to autonomously recognize and react to outside stimuli. Also during our Phase I SBIR, AVNIK developed and conducted a demonstration that allowed 1) a user to query a network system of Apache aircraft fault data and autonomously receive aircraft status; and 2) the intelligent system to autonomously-notify users of new aircraft health status, which provides PHM knowledge to operational/maintenance decision-making at individual aircraft, unit, fleet levels, and sustainment levels. AVNIK also developed statistical anomaly detection algorithms and trend analysis for prognostics prediction algorithms. Our prognostic early warning coupled with CIAs provide supervised autonomous decision making in weapon systems (for such aircraft systems as legacy Apache and upcoming systems, such as FVL) to support operational and ground support maintenance operations.

The overall objective of the FIIN project is to give crew members, maintainers, commanders, and logisticians instant access to information they need to increase safety, maximize system availability, impact Operational & Support costs associated with aircraft maintenance and maintenance-related down time.

The primary question addressed during the study is the feasibility of applying a Cooperative Intelligent Agent (CIA) architecture, operating in a distributed environment with connectivity to a range of maintenance-related resources, to give maintainers automated access to specific information needed to solve maintenance problems.

The work completed by AVNIK has achieved the established objectives through successful completion of the following:

- Assessment of requirements for a Flexible Integrated Intelligent Network (FIIN) and prognostics processing capabilities to be used by Army Aviation operations, sustainment, and engineering organizations.
- Development of FIIN architecture concepts
- Demonstration of selected concepts maintenance-related information resources and maintenance cases from representative platforms
- Evaluation of the development and demonstration results with respect to the requirements.

Presented at the AHS Airworthiness and HUMS Technical Meeting, Huntsville, AL, USA, February 21-22, 2018.

- Results of the study have established the validity of the AVNIK FIIN concept and application of a cooperative agent architecture to integrate Army Aviation enterprise resources. Using a testbed for a FIIN implementation, the AVNIK team has successfully demonstrated retrieval of System Level Embedded Diagnostic (SLED) status information from Apache aircraft by tail number and fault type upon request, and on arrival of new data from recent simulated SLED downlinks. The AVNIK team has also successfully demonstrated prognostics analysis capability using Apache aircraft Maintenance Data Recorder (MDR) parametric data. The demonstrated prognostics analysis capabilities employed anomaly detection and trend analysis algorithms operating on parametric data streams.

CONCEPT DEVELOPEMT

The AVNIK FIIN concept provides an intelligent Prognostic Health Management (PHM) network for Army Aviation based on the new technologies of distributed work environments and CIA systems. The FIIN network will provide means to integrate information collection, distribution and sharing. The network will also provide prognostics analysis and predictive maintenance recommendations both on aircraft platforms and in ground maintenance facilities. This AVNIK concept will lead to a robust network to meet Future Vertical Lift (FVL) systems needs that automates handling and sharing of distributed information from aircraft flight management systems, aircraft and component performance specifications, Condition Based Maintenance plus (CBM+) repositories, and maintenance documentation/resources. The AVNIK concept applies open architectures standards, including consideration of the Joint Common Architecture and Future Airborne Capability Environment (FACE) standards

Intelligent Network

The AVNIK FIIN concept integrates multiple independent associated systems into a virtual system of systems that automates interaction among the associated systems to accomplish joint tasks with minimal operator intervention. This approach employs cooperative intelligent agents to accomplish distributed information sharing functions for combinations of the associated systems and human users, providing dedicated interfaces and performing meta-level processing tasks that involve information from multiple associated systems. The resulting architecture is intended to be flexible and adaptive, maximizing reliability of the network and improving operational availability of the Army Aviation aircraft and support systems in the FVL environment.

Figure 1~~Figure 1~~ is an overall diagram for the proposed FIIN implementation, depicting internal FIIN components and functions, associated system interfaces, and support services/tools that FIIN includes to automate information retrieval and scheduling for unit-level and depot-level maintainers.

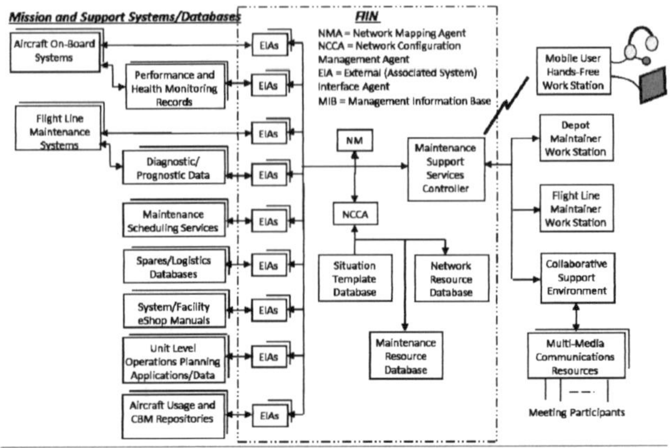

Figure 1 FIIN Conceptual Structure and Interfaces

Control of the overall network is distributed, with the CIAs jointly compiling knowledge of current network configuration/states and working together dynamically to execute adaptive changes for network performance issues that they recognize.

The concept depicted in Figure 1~~Figure 1~~ serves as the executive and/or controller for maintainer support functions, keeping cognizance of real-time inputs from dynamic sources (e.g., aircraft operational status and PHM processes), accepting tasking from users and coordinating presentations of collected information to the users. The network mapping agent (NMA) and the network configuration control agent (NCCA) work together to dynamically determine and control network performance capability.

Use Cases

To define constituents and behaviors in the FIIN Phase I concept, AVNIK developed use cases for three scenarios relevant to Army Aviation PHM capabilities, using Apache PHM data as examples. Two of these use cases are for Apache SLED data processing, one for handling user requests of SLED data, and the other for notifying a user that new SLED data has arrived. The third use case describes a prognostics processing scenario, in which case the processor performed anomaly detection and trend analysis on samples of MDR parametric data.

ARCHITECTURE

The AVNIK FIIN architecture includes design of the agent-based intelligent network, individual agents, and interfaces with users and external associated systems. The architecture also includes design of prognostics processing algorithms and implementation software

Figure 2~~Figure 2~~ is the concept definition and proposed architecture for the FIIN and prognostics use cases and processes.

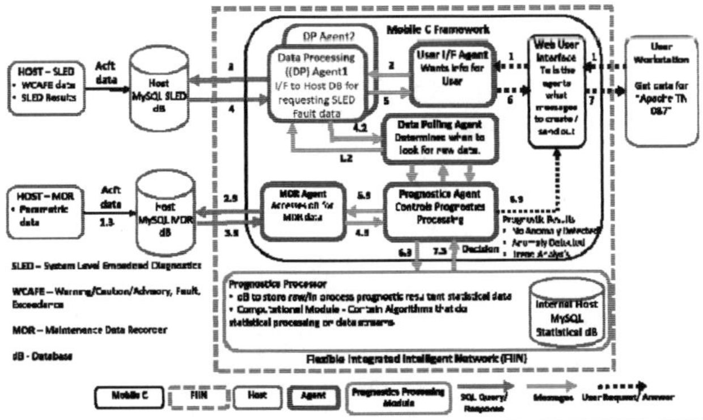

Figure 2 FIIN Concept Definition and Proposed Architecture

Flexible Integrated Intelligent Network

The FIIN cooperative agent architecture uses intelligent software agents to locate relevant knowledge elements and synthesize them to produce a usable information resource. All the agent software structures derive from a single object-oriented base class, currently implemented in C++. Individual agents are objects derived from subclasses of the base class. The individual agents inherit common attributes (i.e., methods or member functions, properties) and knowledge elements used to implement communications with the other agents. Developers create specialized attributes to install in specific subclasses of agents to give agents of those subclasses capabilities to manage specific kinds of information, such as databases, simulations/models, and expert systems. As an example, AVNIK developers programmed Structured Query Language (SQL) query and response methods into agents to request and retrieve SLED and MDR data from MySQL databases via interface with the MySQL database management software.

Each agent incorporates a knowledge base, which contains both static information that developers provide and dynamic information that is received from other agents and external sources during operation. The knowledge contents determine the behaviors of the agent, i.e., the relationships with other agents, non-agent FIIN services, and external systems' interfaces. The agents exchange information with each other and with non-agent resources using a network substructure to pass information and establish interactive sessions as necessary. The primary FIIN maintainer interface employs a web server (anticipated to be a Linux Apache web server), to which maintainers can connect using a standard web browser from any workstation or tablet that has connectivity with the FIIN network.

Individual Cooperative Intelligent Agent Architecture

Figure 3~~Figure 3~~ is a block diagram illustrating the internal software structure of an individual intelligent agent. The agent depicted is representative of a class of agent software objects, from which subclasses and object instantiations having specific functionality and knowledge bases derive. In the diagram, the high-level communications module and the planning-and-coordination module provide functionality for interaction of the agent and the associated system with the network and the outside world. The acquaintance and self-models consist of knowledge bases for context, protocol, and behavior. The associated system interface functionality module and associated interface buffering/storage module provide control and data exchange services as a client of the associated system. The associated system/functionality may connect to an external host through the network or through a dedicated connection/port.

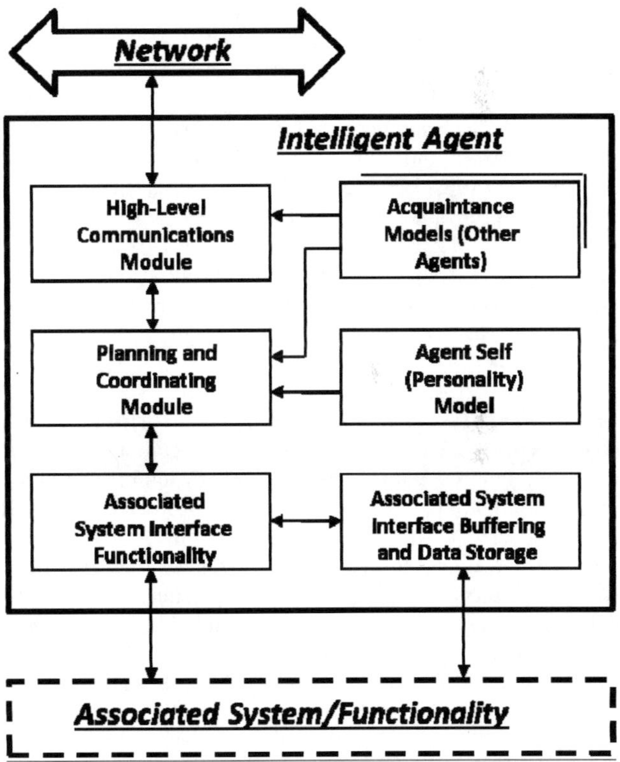

Figure 3 Cooperative Intelligent Agent Architecture

External interface agents (EIAs) connect virtually to the assigned systems and work together to fulfill requests for information from the systems. All the agents are expected to reside within the FIIN boundary (dashed line in Figure 1~~Figure 1~~) unless a given repository or database incorporates an EIA functionality that is either provided by the FIIN or designed/tested to be compatible. The EIAs are specialized knowledge agents, each with rules or procedures installed in a knowledge base for accessing and manipulating contents of the associated system. Where feasible, the EIA rules invoke commercial or standard applications compatible with the respective associated system contents instead of using special-purpose routines. For example, if the target information is in a database that uses SQL commands or requests for access, the EIA may invoke SQL resources stored in an associated system interface functionality module. Engineers or maintainers prepare instruction sets/procedures for complex automated tasks before use when possible, and store the instructions in the situation template database as scripts or macros. Typically, a situation template is activated in response to a FIIN user command or query, or in response to real-time fault information that the controller receives from an associated PHM, diagnostic, or flight management warning function.

The CIA network testbed implementation employs Mobile C, an existing open-source software framework that incorporates agent structures and behavior-specifying capabilities. The AVNIK team enhanced the Mobile C capability to incorporate a C++ wrapper that has the ability to create agents as C++ classes that provide for software reuse and sharing of agent features among agents that have common interfaces and/or behaviors.

The AVNIK team implemented FIIN CIAs as a hierarchy of software classes and subclasses. The top-level agent class contains structures common to all the agents and provides limited sets of static data attributes and operations that enable basic agent behaviors and peer-to-peer agent interactions. Lower level agent classes inherit the top-level class structures and behaviors, and developers add specialized characteristics suited to the types of tasks they will undertake.

The common agent data structures include static characteristics such as the agent's identity (common type but assigned unique values for individual agents and the corresponding situations), the identities of other agents, and host (associated entity) characteristics. In a FIIN, most agents are designed to provide interfaces with other entities such as users, enterprise data processing systems, or database management systems. Agent subclasses implement sets of data attributes and operations appropriate for interactions with those entities. Agents are instantiated at the bottom of the agent class hierarchy as objects for actual use in the FIIN operating network. The instantiated agents must either be initialized with any needed dynamic data values or acquire the values through request operations.

Prognostics

Based on the general prognostics system concepts a FIIN prognostic architecture was developed, consisting of a data conditioning routine, a data similarity testing algorithm, a trending analysis algorithm, and a decision reporting routine. Figure 4~~Figure 4~~ below is a flow diagram of the FIIN prognostics process and the expected implementation of the prototype.

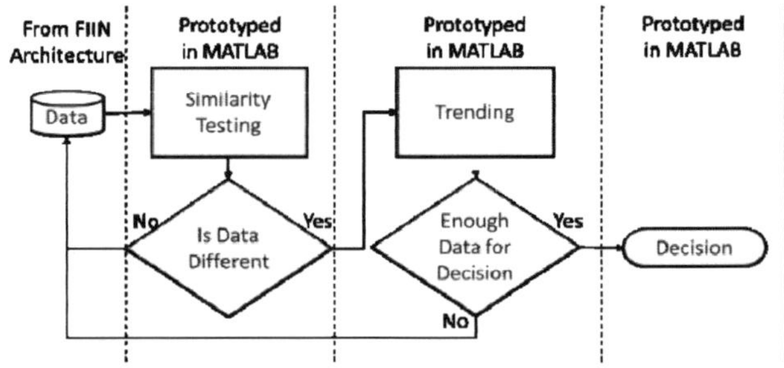

Figure 4 Prognostics Process Flow Diagram

The FIIN data conditioning routine takes the data scheduled for prognostics and ensures that the data is within a defined times series, which the user has specified for when prognostics shall be performed. For the case of this demonstration, data collected during flight was chosen as the appropriate time series for prognostication. The data is conditioned by a logical vector. The logical vector is defined by a condition parameter that the user defines. For example, if the desired data condition is when the aircraft is in flight then the logical vector could be defined as any time samples where the rotor speed is at 100%, or such at time in which the weight is off the landing gear.

The data similarity testing algorithm consists of a two-sample Kolmogorov-Smirnov (KS) test and a Mahalanobis distance calculation. The KS Test provides statistics on the similarity of the test data against a predefined baseline measurement. If the test data is shown to be significantly different from the reference baseline, then a trending analysis of the data is performed and tracked.

The KS test operates on the cumulative distribution of the test data and a predefined baseline. The maximum distance between the two cumulative distributions is calculated. If the distance is greater than a defined critical value then the null hypothesis that the data sets belong to the same distribution is rejected. See Figure 5~~Figure 5~~ below for a graphical depiction.

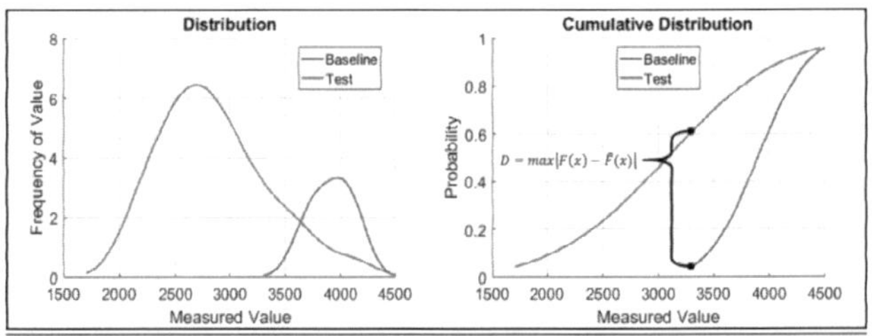

Figure 5 Sample KS Test

In addition to the KS test, a Mahalanobis distance is calculated to give an understanding of how the data is trending overtime with reference to the baseline data. The Mahalanobis distance is used in conjunction with the KS test to discern data similarity or dissimilarity as well as input to the trending algorithm for trending analysis. The Mahalanobis distance is calculated as follows:

$$d(x,y) = \sqrt{(\bar{x} - \bar{y})^T [S]^{-1} (\bar{x} - \bar{y})} \quad (1)$$

Where $\bar{x}$ is the mean of the baseline population, $\bar{y}$ is the mean of the test population, and $[S]^{-1}$ is the inverse combined covariance matrix of x and y. The combined covariance matrix of x and y is calculated as follows:

$$[S] = \frac{N_x [C_x] + N_y [C_y]}{N_x + N_y} \quad (2)$$

Where N_x and N_y are the sample sizes of the baseline and test populations respectively and $[C_x]$ and $[C_y]$ are the covariance matrices of the baseline and test date respectively. The Mahalanobis distance is unitless and scale-invariant, and in the case of this project's application is a measure of how many standard deviations the test data is from the baseline data.

Once the similarity test indicates a dissimilarity, the trending analysis algorithm takes the Mahalanobis distances calculated from the similarity testing algorithm and attempt to fit several modeling functions to the distances to provide a model of how the differing test data is behaving. The modeling functions chosen for this project demonstration were polynomials of degree 1 to 5 and two exponential functions. The following summarizes the models:

$$\begin{aligned} y &= a_1 x + a_2 \\ y &= a_1 x^2 + a_2 x + a_3 \\ y &= a_1 x^3 + a_2 x^2 + a_3 x + a_4 \\ y &= a_1 x^4 + a_2 x^3 + a_3 x^2 + a_4 x + a_5 \\ y &= a_1 x^5 + a_2 x^4 + a_3 x^3 + a_4 x^2 + a_5 x + a_6 \\ y &= c e^{\alpha x} \\ y &= c_1 e^{\alpha_1 x} + c_2 e^{\alpha_2 x} \end{aligned} \quad (3)$$

If a nontolerant limit of the parameter is known, the model can be used to predict when the limit will be reached. Once a model is established, a report is generated for the user to show the results of the trending analysis and estimated time-to-fault scenarios based on trending models of the data. The report will contain the model data and a prediction of when nontolerant limits of parameters will be reached.

RESULTS

A benchtop FIIN engineering prototype was assembled to demonstrate the above automated capabilities. The benchtop FIIN engineering prototype system resides primarily on one or more off-the-shelf workstation computers that operate on a local area intranet. The network uses a commercial router with Ethernet and WIFI ports and will can operate as a subnet on a larger network. A portion of the FIIN prototype system that simulates an on-platform prognostics processing element utilizes a dedicated processor unit capable of being upgraded to aircraft physical operating environments. At least initially, the FIIN prototype network was kept isolated from the internet.

To the extent feasible, the prototype system uses open-source software, including Linux operating systems, MySQL or PostgreSQL database management systems, and cooperative agent development framework applications, such as the Mobile-C agent framework. Because some of the potential host systems that interface with the FIIN may use Microsoft operating systems, a plan to incorporate a windows workstation in the prototype network to provide resources for testing agent interfaces to Windows applications/operating systems was planned for future development.

Intelligent Network

The FIIN benchtop demonstration implemented the two SLED data handling scenarios, as described above. Scenario 1 demonstrated the process that the CIAs follow to service a user request for fault data and other status information for a given aircraft tail number. Scenario 2 demonstrated the response to arrival of new SLED data, which results in a notification of the existence of the new data to the user. In both cases the SLED data was preloaded into the benchtop test MySQL database to be retrieved upon queries by agents, simulating behavior of external host system data sources.

For the FIIN demonstration, the primary FIIN interfaces were with the user and outside information sources (i.e., resources accessible over the Internet or enterprise intranets, such as the SLED database).

The FIIN user interface consisted of a web-based graphical user interface (GUI) implemented via a web-browser for ease of user access to the FIIN system from remote locations. We implemented the web-server functionality using a Linux Apache Web-Server. Users of the FIIN at remote terminals may access the web-server using an off-the-shelf or open source browser.

Using the open-source Mobile-C multi-CIA development framework with a C++ "wrapper", several classes of CIAs were created to serve required functions such as interfacing with the MySQL database server, interfacing with human users, and performing testbed housekeeping functions. The Mobile-C framework instantiates individual CIA objects as members of the CIA classes as needed. The user interface consists of a Linux Apache web server, which connects to a web browser at the user workstation. The Data Polling CIA and the Data Processing CIA depicted above (Figure 2~~Figure 2~~) communicate with the User Interface CIA, which also interfaces with the web server. The Data Processing CIA connects to the data source interface.

Figure 6~~Figure 6~~ below is a screen shot of the Phase I Scenario 2 display depicting the user interface for the FIIN demonstration. As noted for Scenario 1, the screen presents time-stamped data for a specific Tail Number and Fault data. If the operator had selected a Fault Id Search a different output screen would be displayed. This screen shot also shows the new data displayed in response to the operator clicking the green button in the upper right-hand corner of the window.

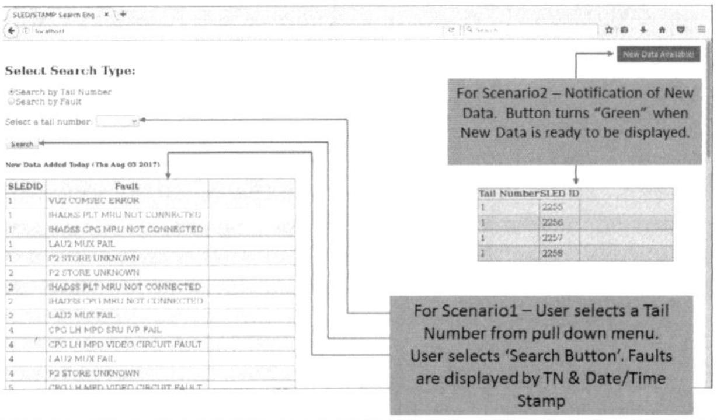

Figure 6 FIIN Architecture – User Interface

Prognostics

Apache PM supplied MDR data for developing and testing the FIIN prognostics process. Parameters were chosen from the MDR data to characterize the aircraft flight status, so that the data was processed only during flight. Engine oil pressure, rotor speed, and landing gear load parameters were chosen to determine time windows where the aircraft was in flight. A logic vector was implemented by windowing all time samples where either engine oil pressure exceeded 20 psi and the rotor speed was greater than 0. The logic vector was used to condition the raw test data so that prognostics was performed only when the aircraft was in flight.

Several potential data sets were chosen for FIIN verification testing of prognostics, including: Battery Voltage, AC Electrical Busses, DC Electrical Busses, Transmission Temperature, Transmission Pressure, and Hydraulic Pressure.

For this prognostics verification testing, the Battery Bus Voltage was chosen since the data was consistent throughout the entire flight. Several assumptions were made for the purposes of verifying FIIN prognostics. The Battery Bus voltage at 28 volts was assumed to be a normal working condition for the aircraft. A Battery Bus Voltage of 26 volts was assumed to be a nontolerant fault that would cause damage to the system.

A seeded fault was implemented into a sample of the Battery Bus Voltage data by subtracting a decreasing trend simulating a fault in the DC Battery cells. The trend started at sample number 52068. The nontolerant limit of 26 volts would be reached at about sample 65000. The goal of the prognostics system was to correctly determine a fault had occurred shortly after the fault occurred and accurately predict when the fault would cause the system to fail.

The following is a sample output of the prognostics process algorithm (Figure 7~~Figure 7~~ – Figure 11~~Figure 11~~). The purple dots represent the window of data points that are used as test data. The yellow dot is where the fault was initiated in the raw data. Figure 7~~Figure 7~~ is the first windowed test of the data. Figure 8~~Figure 8~~ shows the output shortly before the fault starts. The algorithm is operating as expected indicating no fault has occurred. Figure 9~~Figure 9~~ shows the output one sample after the fault occurs. The algorithm correctly detected a fault 0.1 seconds after the fault occurred. A new line in the middle graph shows the current prediction model. The model at this point has only 2 samples to create a prediction. The algorithm is operating as expected by fitting a line to these two points. Figure 10~~Figure 10~~ shows the sample output 3000 samples after detecting a fault (294 seconds). With 3000 data points the algorithm is outputting a reasonable trend predicting the nontolerant fault limit will occur in 779 seconds when the actual limit will be reached in 1141 seconds. With more iterations, the prediction model increases in accuracy (Figure 11~~Figure 11~~).

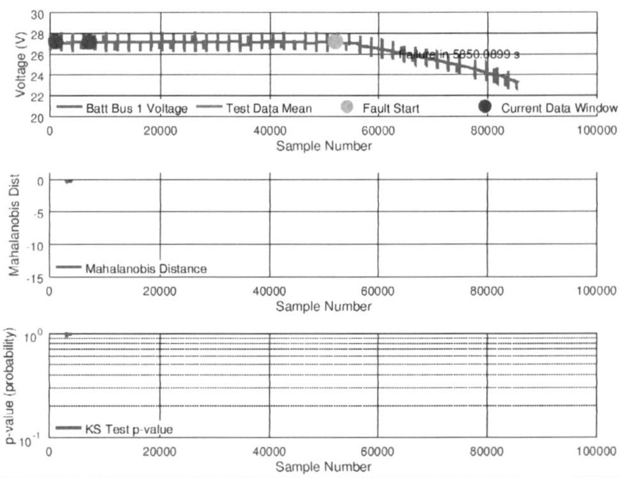

Figure 7 Prognostic Process Sample Output (Frame 1)

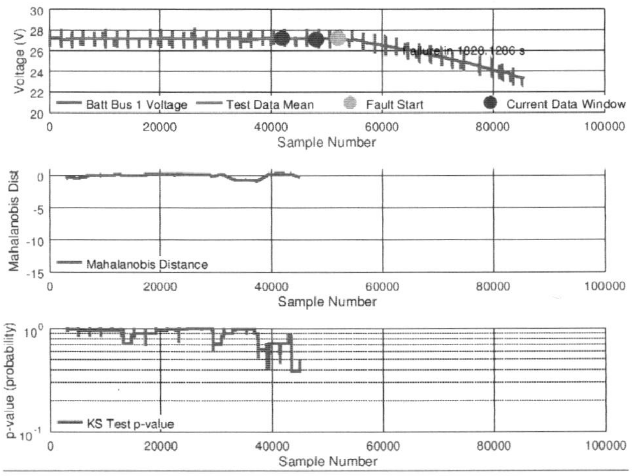

Figure 8 Prognostic Process Sample Output (Frame 46)

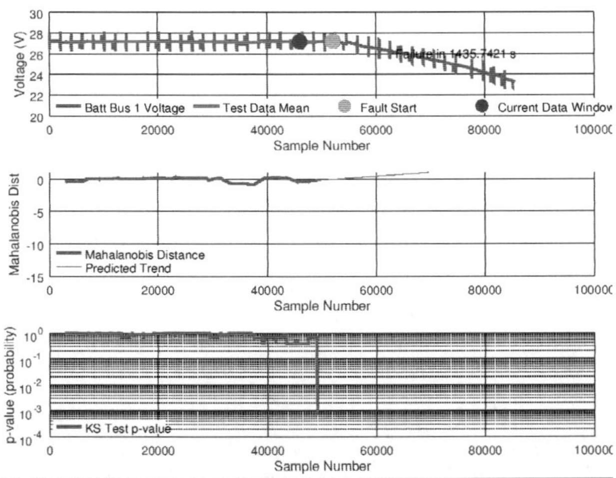

Figure 9 Prognostic Process Sample Output (Frame 48)

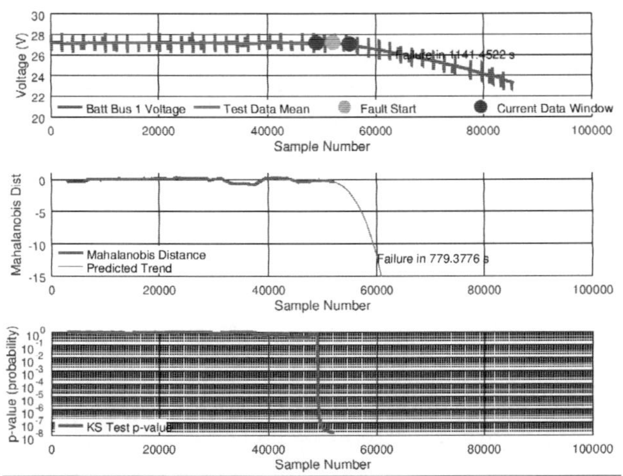

Figure 10 Prognostic Process Sample Output (Frame 51)

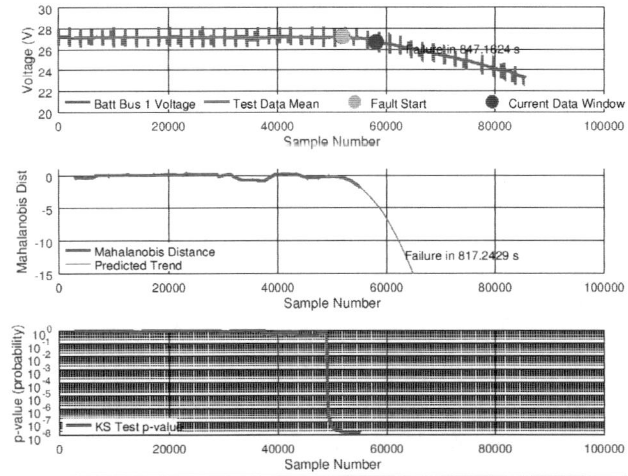

Figure 11 Prognostic Process Sample Output (Frame 54)

CONCLUSIONS

In this study, the feasibility and potential benefits developing an automated robust FIIN for PHM to support Army Aviation mission operations, maintenance, and logistics processes, using cooperative intelligent agent technology for acquisition, processing, and distribution of aircraft systems health and status information was shown. During the study, a class of CIAs were designed and object-oriented software design methodology were used to create subclasses of CIAs specialized to perform user interface tasks, to work with database management systems, and to coordinate data flow to/from prognostics analysis algorithms. The CIAs interpreted user requests and host system information elements through a coordinated set of ontologies adapted from aircraft maintenance technology and languages used by the associated systems (e.g, SQL for database management systems). Prognostics processing algorithms and data flow management structures were adopted from a previous AVNIK structural health monitoring (SHM) project sponsored by Army Aviation and Missile Research Development Engineering Center (AMRDEC) and a previous ISI predictive fault detection project sponsored by the Missile Defense Agency (MDA).

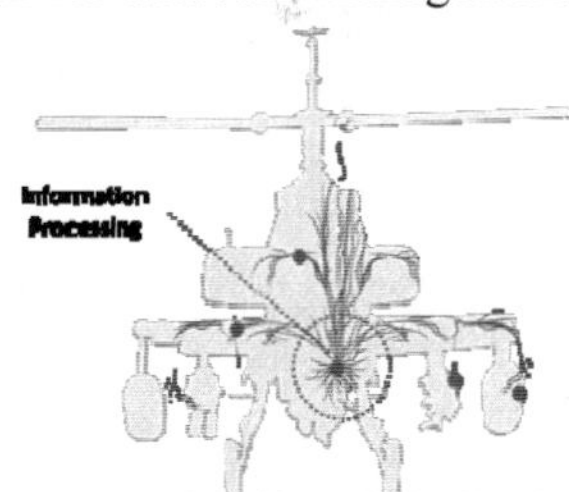

Structural Health Monitoring (SHM)

The FIIN study demonstrated a successful integration of data from multiple sources into a complex intelligent agent system of systems network that provided a unique and novel technique to display maintenance data and provide prognostic capability to alert maintainers and crew of potential actionable events.

Author contact:
W. Thomas Little thomas.little@avnikdefense.com
Michele K Platt michele.platt@avnikdefense.com
Jeff Yalowitz yalowitz@insciences.com

ACKNOWLEDGMENTS

AVNIK acknowledges the US AMRDEC Advanced Development Directorate (ADD) for funding this Phase I SBIR effort and to Mr. John Moffatt for his support as technical monitor. AVNIK also acknowledges Apache Program Management Office (PMO), Mr. Brian Murphy and AMRDEC Software Engineering Directorate (SED) for supporting the FIIN for PHM Phase I SBIR by providing actual AH-64 SLED and MDR data samples for processing and demonstration.

REFERENCES

[1] R. K. Youree, J. S. Yalowitz, et al, ìA Multivariate Statistical Analysis Technique for On-Line Fault Predictionî, *International Conference on Prognostics and Health Management*, 2008

[2] W T Little, et al., "Optimized Structural Health Monitoring System Design for Aviation Structures," in Proc. From *11th International Workshop on Structural Health Monitoring.*, Stanford, CA, 2017

STANDARD FLEXIBLE INTEGRATED INTELLIGENT NETWORK (FIIN) ARCHITECTURE FOR INTEGRATED VEHICLE HEALTH MANAGEMENT (IVHM) SYSTEMS

February 21, 2018

AHS Airworthiness, CBM & HUMS Technical Meeting, Huntsville, AL

GLOBAL STRATEGIC SOLUTIONS

Content Outline

- Motivation
- Current State
- Architecture Development Approach
- Solution
- Architecture Instantiation & Standardization Path
- Takeaways/Future Work
- Discussion

GLOBAL STRATEGIC SOLUTIONS

Motivation

- ***Nine (9) core system objectives were derived from the Army ADD/AATD FIIN topic description:***
  1. Provide autonomous IVHM data acquisition (DA), data manipulation (DM) and exchange with onboard health-ready systems;
  2. Provide a means for configuring and programming IVHM data acquisition from specific health-ready systems;
  3. Provide IVHM data connectivity to multiple onboard data buses;
  4. Provide connectivity to vehicle safety interlock functions;
  5. Provide for reliable, real-time deterministic processing and transmission of IVHM data;
  6. Provide intelligent self-monitoring and self-correction;
  7. Provide for processing of cross-domain IVHM data;
  8. Provide state detection (SD) and health assessment (HA) for health-ready components, and;
  9. Provide a modular, open-systems, standards-based architecture.

GSS
GLOBAL STRATEGIC SOLUTIONS

 GSSLLC (C) 2018. ALL RIGHTS RESERVED

Current State

GSS
GLOBAL STRATEGIC SOLUTIONS

Current State in IVHM/CBM+

- **Current state** based on recent study conducted by GSS and LMI for the Director, Enterprise Maintenance Technology, Office of Deputy Assistant Secretary of Defense, Maintenance Policy and Programs (ODASD MP&P); *"CBM+/IVHM System Interoperability and Standards"*
- **CAPABILITY LEVELS FOR AEROSPACE & AUTOMOTIVE IVHM APPLICATIONS,** *JA6268, "Design & Run-Time Information Exchange for Health-Ready Components"*
- Given the current state, ***FIIN is a foundational technology*** *that could enable higher IVHM capability levels in Defense aviation platforms.*

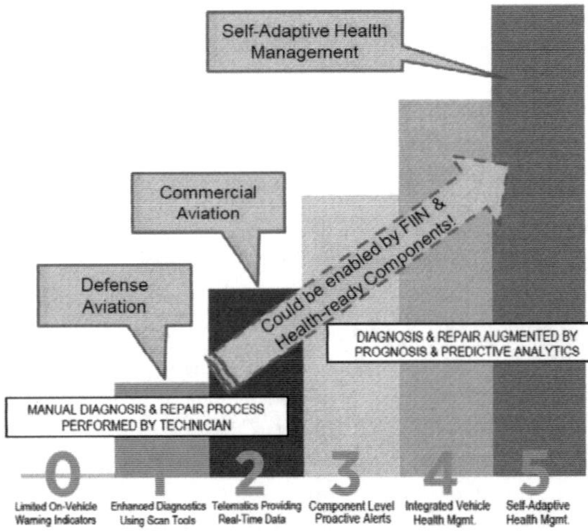

GSS GLOBAL STRATEGIC SOLUTIONS

Best Practice in Commercial Aviation

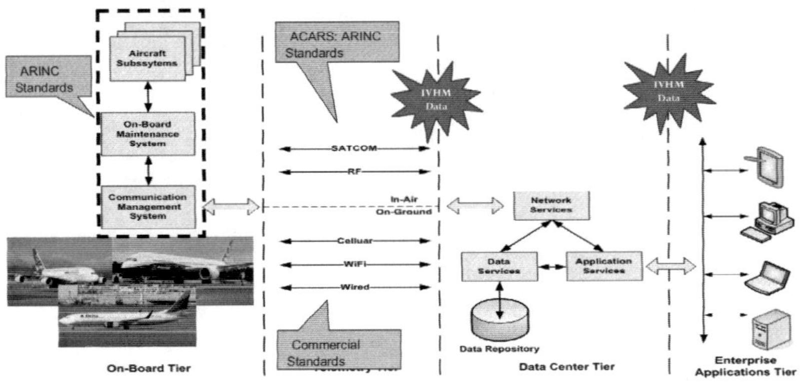

- ✓ Standards-based centralized systems gather health and status data from several subsystems and perform fault correlation and root-cause analysis, directing the mechanic to the subsystem that requires repair or replacement.
- ✓ Data collected; BIT, FOQA, aircraft performance, (e.g., hard landings, engine, performance, tail strikes, etc.).
- ✓ Some OEMs offer health monitoring/advisory and analytics services for their equipment as part of their product support package.

GSS GLOBAL STRATEGIC SOLUTIONS

IVHM System Capabilities in Defense Aviation

IVHM System Functional Block	Platform	Data Transfer Tier (Operational)	Ground Support	Data Transfer Tier (Enterprise)	Enterprise
Advisory Generation (AG)	Subsystem BIT Faults & ICAWS Indicators		Subsystem BIT Faults & ICAWS Indicators		
Prognostics Assessment (PA)	Life parts tracking, overload/regime recognition		Structures and mechanical parts life management		Structures and mechanical parts life management
Health Assessment (HA)	BIT filtering and fault correlation		BIT correlation to fault isolation maintenance procedures /IETM		
State Detection (SD)	System BIT, enhanced diagnostics, exceedance monitors (e.g., HUMS)		Initiated BIT, HUMS condition indicators processing		
Data Manipulation (DM)	Done within system BIT, HUMS/IVHM processor, etc.		Automated Logistics Environment, HUMS ground station, PMA		CBM Data Repository
Data Acquisition (DA)	Subsystem sensors, HUMS/IVHM processor, data recorder	Data is downloaded via DTU, PMA, data recorder (telemetry for Triton/Apache/JSF)		Data can be transmitted via JDMS to enterprise data repository or cloud service	

GSS
GLOBAL STRATEGIC SOLUTIONS

 GSSLLC (C) 2018. ALL RIGHTS RESERVED

Architecture Development Approach

GSS
GLOBAL STRATEGIC SOLUTIONS

Why a Standards-based FIIN Architecture?

Current challenges in Defense applications found in OSD study:

- **Data Ownership**:
  - OEM dependence & data ownership issues.
  - Programs and Maintenance organizations limit access to their data.
  - "Perceived value" and "classification" of raw asset data maintenance information limit rapid dissemination across the enterprise.
- **Connectivity***:
  - Non-integrated IVHM data supply chain.
  - At/off-platform test and maintenance information is still captured into paper records or within isolated databases.
  - Ground maintenance and logistics information systems may limit the ability of users to extract relevant data in a form that can be communicated to other such system.
- **Data Quality***:
  - Dissimilar preservation of data source context (Values, Meta Data)
  - The data collected across operations and maintenance activities is in numerous non-standard formats

*Related to non-integrated on-board/off-board IVHM architecture

GSS
GLOBAL STRATEGIC SOLUTIONS

Modular Open Systems Approach

FIIN Objective

Develop and demonstrate an integrated, robust, flexible, and intelligent network for Army aviation IVHM applications

Design Principles

- Establish Network Parameters
- Employ Modular Design
- Designate Key Interfaces
- Select Open Standards
- Document FIIN Architecture

Target FIIN Benefits

- Ease of Change & Integration
- Reduced Development Cycle Time & Total Ownership Cost
- Architecture Re-configurability
- Life-long Supportability
- Continued Access to Cutting Edge Technologies
- Commonality & Reuse

Modular Open Systems Approach (MOSA)

Characteristics of IVHM Systems:

- Evolving technologies and system(s)
- Cost/benefit driven
- Multiple contributors (enterprise, users, OEM aircraft/subsystems, technology suppliers, safety,...)
- Major portion of development & maturation post flight test during operational deployment
- Continuous updating in a mixed criticality environment (flight/mission/support critical elements)
- Potential to leverage commercial technology base and applications

GSS
GLOBAL STRATEGIC SOLUTIONS

FIIN IVHM Architecture Elements

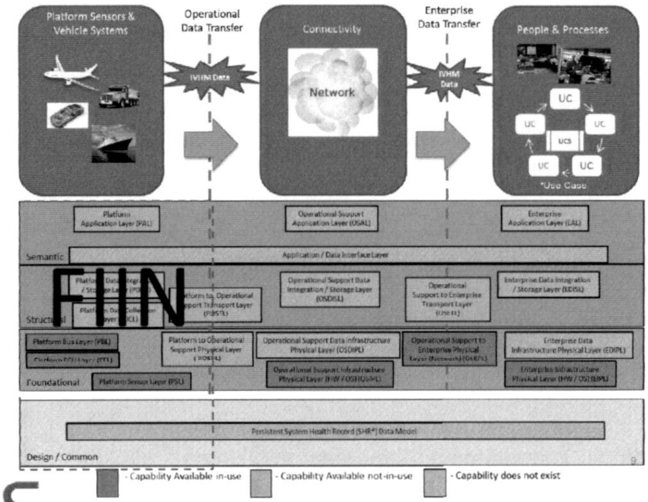

GSS
GLOBAL STRATEGIC SOLUTIONS

Development Approach

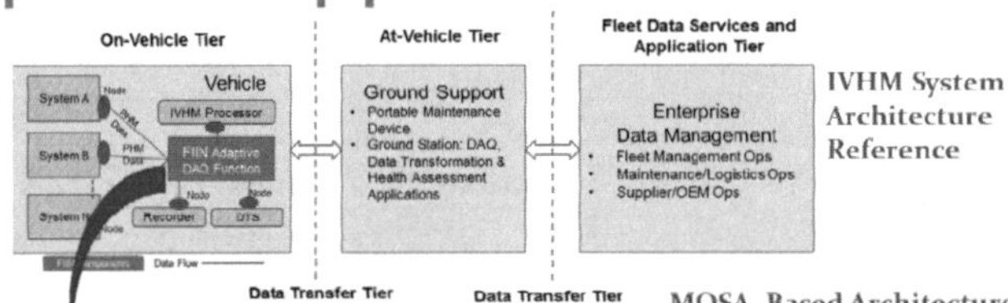

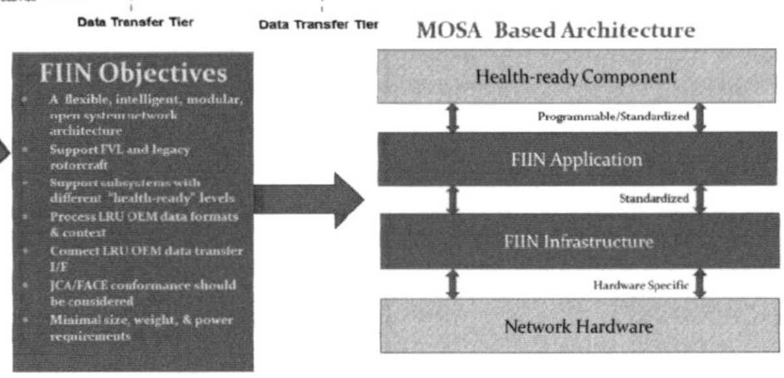

- **FIIN Architecture Focus**: integrate rotorcraft On-board IVHM functions in new and legacy platforms.
- **Approach**: hardware and software independence.

GSS
GLOBAL STRATEGIC SOLUTIONS

Solution

GSS
GLOBAL STRATEGIC SOLUTIONS

The Solution

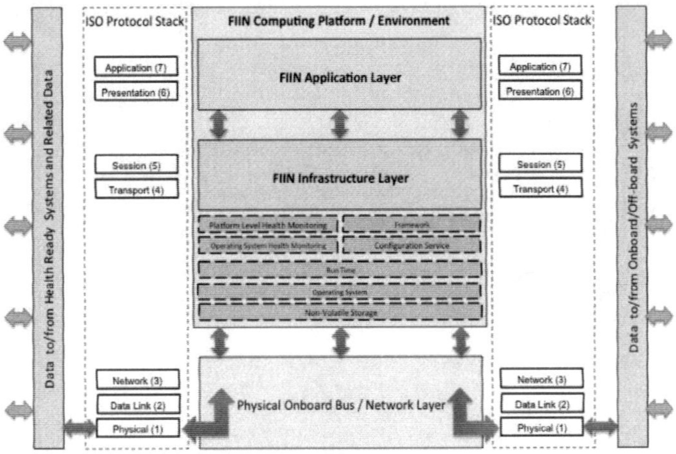

GSS
GLOBAL STRATEGIC SOLUTIONS

FIIN Architecture Benefits

- Designed to be robust, flexible, with the ability to reconfigure / self-heal, and to ensure network integrity, accuracy and reliability.
- Provides machine intelligence and a high degree of automation.
- Enhances local situational awareness by integrating diagnostics, warnings and other IVHM data.
- Reduces the operational burden on users, specifically the aircrew and the maintainers.
- Enables cost-conscious integration of IVHM capabilities into future and legacy Army rotorcraft.
- Provides a "plug and play" architecture versus the traditional "bolt-on" integration approach.
- Facilitates the reuse of software components across multiple platforms
- Advances CBM+/IVHM system interoperability and standards-by facilitating data exchange from the vehicle to the ground support and enterprise tiers

GSS
GLOBAL STRATEGIC SOLUTIONS

Architecture Instantiation & Standardization Path

GSS
GLOBAL STRATEGIC SOLUTIONS

FIIN Architecture Instantiation Elements

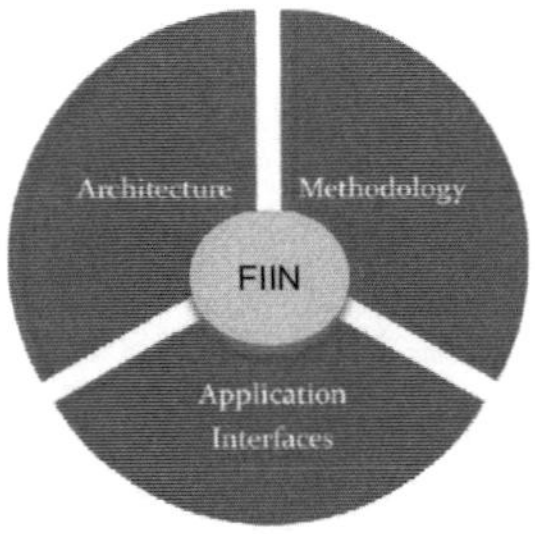

- **Architecture:** Software architecture including a the software stack for the data communications network– the FIIN Infrastructure Software – as an integration platform for hardware independent applications.
- **Methodology:** Standard resource description templates to enable a seamless configuration process of the infrastructure and the integration of application software in health-ready components and the methodology for using this framework.
- **Application Interfaces:** Specification of interfaces of typical applications from all domains (Propulsion, Drive Systems, Electrical Systems, Structures, Rotors & Dynamic Components, Flight Controls & Hydraulics) in terms of syntax and semantics, which should serve as a standard for health-ready component application software.

GSS
GLOBAL STRATEGIC SOLUTIONS

FIIN Architecture Instantiation (Notional)

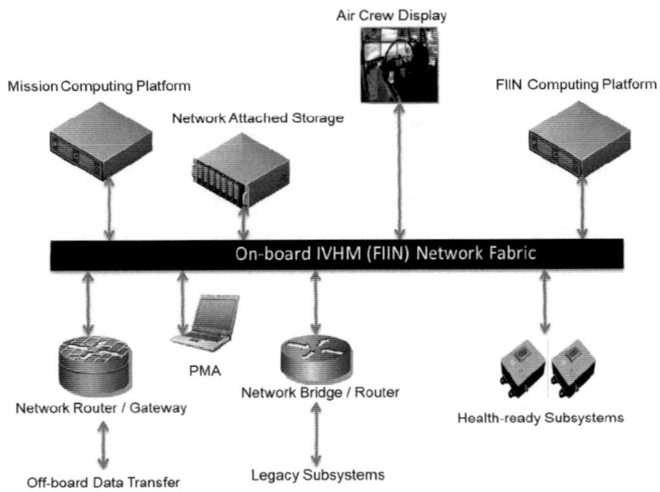

Note: Leveraging ARINC 664 and 825 and other existing standards in the FIIN network fabric architecture can move FIIN towards a "standardized open system framework" for the on-vehicle IVHM architecture tier.

GSS
GLOBAL STRATEGIC SOLUTIONS

FIIN Standardization Roadmap (Notional)

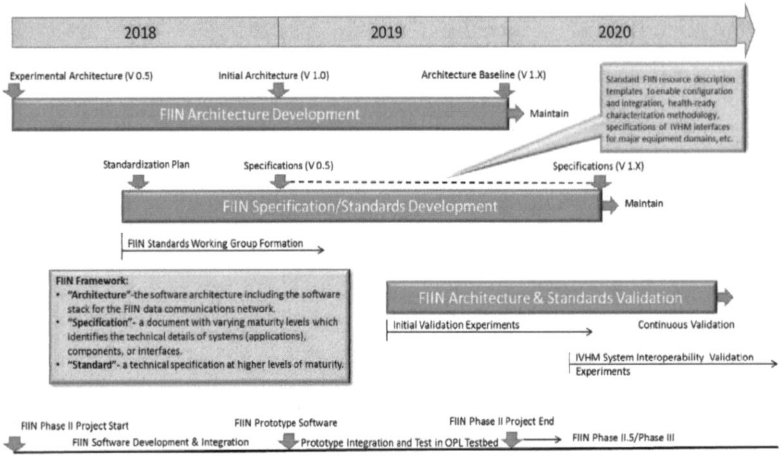

GSS
GLOBAL STRATEGIC SOLUTIONS

Takeaways & Future Work

Takeaways:

- The FIIN architecture concept was developed considering (a) the current state of IVHM in aviation and (b) future platforms (e.g. FVL) where there might still be a need to support connectivity to "legacy" style buses.
- In the case of legacy rotorcraft, FIIN can simplify on-board data collection and processing, but the use of open standards would be more limited.
- A design instantiation of the FIIN architecture can be done on a legacy platform to advance the legacy platform IVHM capability and move FIIN toward a standardized open system framework.

Future work:

- Develop, integrate and test the core services and capabilities in the FIIN architecture in a systems integration laboratory (SIL) environment.
- Actual integration and test in commercial rotorcraft platform equipped with HUMS & health-ready subsystems, in-flight IVHM data collection.

GSS
GLOBAL STRATEGIC SOLUTIONS

Discussion

Contact Information:
Luis Hernandez
Managing Director
703-466-0500, X2002
luis@gssllc.net

GSS
GLOBAL STRATEGIC SOLUTIONS

Estimation and Control of Structural Loads and Stresses in Fatigue Critical Components

J.V.R. Prasad **Julian J. Rimoli** **Massimo Ruzzene**
School of Aerospace Engineering
Georgia Institute of Technology – Atlanta GA 30332

ABSTRACT

Current methods for structural health and usage monitoring and load alleviation control rely on distributed sensing and operational monitoring to infer usage and estimate fatigue in critical components. Such inference process is affected by significant uncertainty given that sensors' type and locations are often removed from hot spot areas characterized by maximum stresses. For example, past work for limiting pitch link loads has used proxy models of the vibratory loading. Recent studies have demonstrated the potential for limiting the peak-to-peak pitch link loads by limiting roll rate, thereby reducing incremental fatigue damage. Also, methods that approximate coupled body/rotor dynamics using high order Linear Time Invariant (LTI) models employ harmonic decomposition to represent higher frequency harmonics as states, and offer the potential for real-time estimation of the effect of control inputs on component dynamic loads. These in turn can be used in combination with reduced order structural models to estimate primary damage variables associated with fatigue of critical components. Such real-time estimation of component level dynamic loads, stresses and strains, etc., provides the opportunity for real-time monitoring of component damage variables, and the development of control schemes designed to alleviate component fatigue damage. The paper describes ongoing efforts towards the development of real-time algorithms for estimation of component level dynamic loads, which enable the evaluation of component level stresses, strains and primary damage variables. These estimations are employed for the development of load alleviation and damage mitigating control schemes for critical component life extension with little impact on handling qualities. The paper will provide a description of the methods being pursed with notable findings to-date.

INTRODUCTION

A 2012 survey of the past 30 years, carried out within Augusta Westland Limited (AWL) Materials Technology Laboratory, concluded that fatigue failures account for approximately 55% of all premature failures in helicopter components[1]. The causes of low cycle fatigue are largely due to aircraft maneuvers, gust loading and through takeoff and landing. Critical helicopter components, classified as Grade-A Vital components by regulatory authorities, are subject to significant fatigue loading in which the failure would result in a catastrophic event. A list of fatigue critical components[2] on the AH-64A Apache shows that many of the Grade-A Vital components are located in the rotor system, creating challenges for real time load monitoring of those components.

Current methods for structural health and usage monitoring and load alleviation control rely on distributed sensing and operational monitoring to infer usage and estimate fatigue in critical components. Such inference process is affected by significant uncertainty given that sensors' type and locations are often removed from hot spot areas characterized by maximum stresses. For example, past work for limiting pitch link loads has used proxy models of the vibratory loading. A classic example is the Equivalent Retreating Indicated Tip Speed (ERITS) parameter, which has been correlated as a function of airspeed and normal load factor with vibratory pitch link loads from retreating blade stall onset, can be limited to indirectly constrain the pitch link loads.

A recent Penn State study[3] (Figure 1), using curve fits of pitch link vibratory loads as function of aircraft states, demonstrated the potential for limiting the peak-to-peak pitch link loads ('with DMC' in the figure) by limiting roll rate in a high fidelity FLIGHTLAB simulation of a utility helicopter, thereby reducing incremental fatigue damage (as measured by a crack growth model), in this case by over 50% for a roll reversal maneuver. The role rate limiting resulted in no degradation in handling qualities when evaluated using the ADS-33 attitude quickness and bandwidth specifications. A recent Georgia Tech study[4] has developed methods to approximate coupled body/rotor dynamics using high order Linear Time Invariant (LTI) models. These methods use harmonic decomposition to represent higher frequency harmonics as states in an LTI state space model, and they offer the potential for real-time estimation of the effect of control inputs on component dynamic loads, which in turn can be used in combination with reduced order structural models to estimate primary damage variables associated with fatigue of critical components. Such real-time estimation of component level dynamic loads, stresses and strains, etc., provides the opportunity for real-time monitoring of component damage variables, and the development of control schemes designed

Presented at the AHS Airworthiness and HUMS Technical Meeting, Huntsville, AL, Feb 21-22, 2018.

to alleviate component fatigue damage. The key issues are estimation fidelity, computational efficiency and controller robustness.

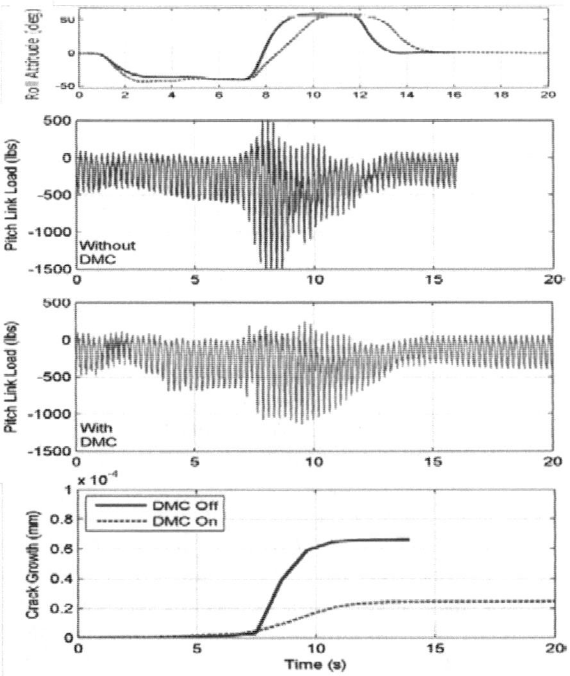

Figure 1. Curve fits of pitch link vibratory loads as function of aircraft states demonstrating the potential for limiting the peak-to-peak pitch link loads by limiting roll rate (taken from Ref. 3).

The paper describes an ongoing study whose primary objectives are the development of real-time algorithms for estimation of component level dynamic loads, along with real-time algorithms for estimation of component level stresses, strains and primary damage variables. These algorithms will be then employed for the formulation of load alleviation and damage mitigating control schemes for critical component life extension with little impact on vehicle handling qualities. Finally, the study will consist in evaluations of the developed algorithms and control designs through batch and piloted simulations.

TECHNICAL APPROACH

The overall technical approach encompasses the detailed prediction of aerodynamic loads through FLIGHTLAB simulations. These loads are then fed to a detailed Finite Element structural model of a dynamic component of interest. In this case, the pitch link is considered as representative element employed in the initial investigations of this project. The structural model is used to compute the dynamic response resulting from the application of the aerodynamic loads, which provides the harmonic content of the dynamic structural response of the component. These harmonic terms, and the corresponding stresses, are then fedback as inputs to the control algorithm that implements the load mitigation procedure to be developed in this work. A high-level schematic of this approach is presented in Figure 2.

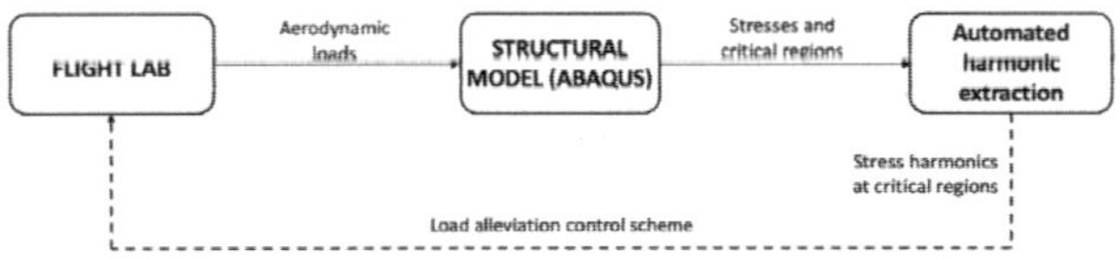

Figure 2. Schematic of technical approach to the development of load mitigation control algorithms

The structural models will be formulated here to account for the presence of stress concentration areas, and the local presence of damage. To this end, the computational cost of the Finite Element Method (FEM), which can become prohibitive due to the presence of small components relative to a large simulation domain is addressed through model order reductions that rely on the formulation of multi-scale numerical schemes. The proposed computational framework will provide tools to estimate the effect on life and damage progression, and potentially support the formulation of control algorithms that extend operational life and increase safety in the presence of damaged components.

Details of the overall technical approach are presented in the following sections.

Component Load Estimation

Our approach to developing real-time algorithms for estimation of component level dynamic loads makes use of high fidelity nonlinear model of a notional conventional rotorcraft (e.g. representative of UH-60) in FLIGHTLAB. The model will include flexible blades, flexible body and flexible control system linkage models, and thus be suitable for calculation of dynamic loads. We will develop LTI models of coupled body/rotor dynamics across the envelop, including quasi-steady flight conditions such as pull-ups and turns using the methods we have developed for LTI model extraction from nonlinear models[4]. As the dimension of the LTI models can become very large depending on the number of harmonic states retained, we will investigate the use of balanced realization schemes for reducing the LTI model order suitable for real-time implementation. We will evaluate the fidelity of the reduced order LTI model through comparisons with the nonlinear model in terms of error response plots in the frequency domain using CIFER. While the reduced order LTI models can be used for real-time component load estimation, the estimated loads may have to be corrected for errors due to LTI model approximations, nonlinearities, etc. We will construct a Linear Quadratic Estimator (Kalman Filter) using fixed system load measurements to correct LTI model state estimates. The proposed LTI/LQE system for component load estimation is shown in Figure 3.. It is possible that some of the LTI model states that are important in component load estimation become unobservable for a selected set of fixed system measurements. We will address this issue using *Observability Gramians* to determine the minimum number of sensors and their locations (including, if needed, limited sensing in the rotating system such as blade flapping) needed for LTI model state estimation. We will verify the performance of the proposed LTI/LQE system through extensive nonlinear model simulations of selected maneuvers with simulated measurement noise and external gust disturbances (Figure 3). We will also validate the proposed LTI/LQE system using available component loads data from flight testing.

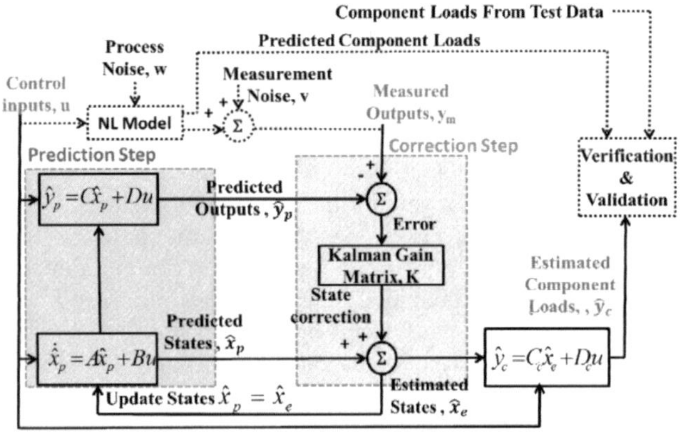

Figure 3. Verification and validation of LTI/LQE system for component load estimation

Reduced Order Component Structural Models

The Finite Element Method (FEM) is generally the preferred choice for modeling problems involving dynamic loads, mainly because of its flexibility to represent arbitrary geometric features. In the case of structural health monitoring and life extending control of rotorcraft, however, the computational cost of using FEM can become prohibitive due to the presence of small components (e.g., pitch horns) relative to a large simulation domain (e.g., main rotor system assembly). This difficulty arises from the fact a large number of small elements are required to capture the geometry of small components, which, in turn, results in large problem sizes with small stable time increment for temporal integration. Thus, successful simulation of such structural systems at reasonable computational cost requires the adoption of reduced-order simulation techniques. One approach to model order reduction is through the formulation of multi-scale numerical schemes. In our approach, we will implement an extended version of the Geometric Multiscale Finite Element Method (GMsFEM)[5] formulation to better predict the dynamic response of multi-scale structures. By enriching the set of GMsFEM shape functions with a number of vibration normal modes, the model will utilize macro-elements to resolve sub-element wavelengths, which in turn will help avoiding excessively small stable time increments usually incurred by the presence of small geometric features. To achieve this, we will divide finite element interpolation functions into two categories: boundary shape functions $\boldsymbol{h}$ and interior shape functions $\overline{\boldsymbol{h}}$, resulting in an interpolated solution $\hat{u}$ for the displacement field given by $\hat{u}(x,t) = \boldsymbol{h}(x)u(t) + \overline{\boldsymbol{h}}(x)\bar{u}(t)$. In the previous expression, $u(t)$ are traditional nodal (displacement) degrees of freedom, whereas $\bar{u}$ are degrees of freedom associated to a particular vibration mode. That is, functions $\boldsymbol{h}(\boldsymbol{x})$ are traditional finite element interpolation shape functions able to capture the quasi-static deformation of multi-scale elements, whereas functions $\overline{\boldsymbol{h}}$(x) are the natural vibration modes of the element subject to fixed boundary conditions, allowing us to capture the sub-element dynamic response in a computationally efficient manner. The natural vibration modes are pre-computed using an auxiliary sub-element grid.

Once the extended GMsFEM method has been implemented, the resulting reduced order (or multi-scale) model will provide, in real-time, detailed information of the dynamic stress distribution within mechanical components of interest. These stresses, in turn, will be evaluated according to various damage accumulation criteria at the finite element quadrature point level as follows. First, we will assume an internal flaw and flaw size distribution for each component. If stresses are such that no plastic strains are generated, a high cycle fatigue criterion, e.g. Paris Law, will be point-wise adopted. Through this criterion, growth rates for assumed internal flaws will be evaluated. Thus, a state variable L, representing current flaw size, will be stored at each quadrature point of the enriched GMsFEM formulation. If stresses are such that plastic strains are generated, a continuum damage evolution criterion, e.g. Johnson-Cook with damage, will be adopted. In this case, the local state variable D, representing accumulated damage, evolves proportionally to the rate of plastic strain. Each of these criteria are known to produce reliable damage/degradation estimates within their range of applicability. However, the situation is not entirely clear when a component experiences both kinds of loads during its operational life. A standard engineering practice to account for this complex scenario is to estimate life through phenomenological criteria, e.g. those given by the Soderberg/Goodman type of curves. Part of this research effort will be aimed at extending these phenomenological approaches by exploring the possibility of including mechanistic behaviors into models accounting for combined high cycle fatigue and plastic damage. That is, we propose to develop a formulation based on a combined accumulated damage estimator $D_{comb}(L, D)$ to predict the life of a component. For example, when there is no plastic strain, a measure associated to the fatigue damage could be $D_{fat} = L/L_{cr}$ if $L < L_{cr}$ and $D_{fat} = 1$ if $L > L_{cr}$, where L_{cr} is a critical flaw size that would induce component failure. On the other hand, in the case where plastic strains are present, the plastic damage variable results to be the same as for the Johnson-Cook model, i.e., $D_{plastic} = D$. A logical approach to account for cases where both mechanisms are present could be to adopt a convex combination of both parameters, which resembles Soderberg's approach. We propose to adopt a combined damage variable given by $D_{comb} =$

$\alpha\, D_{fat} + (1-\alpha)\, D_{plastic}$, where α is a constant to be determined (a first attempt could be to adopt $\alpha = 1/2$). In this way, our reduced order model will generate a combined damage estimation D_{comb} which, in turn, will serve as the control variable for component life extension.

Life Extending Control Schemes

A control strategy for component life extension is to reduce component dynamic (for example, peak-to-peak) loads, leading to reduced peak-to-peak stresses, and hence, potentially leading to reduced fatigue life usage. In this Load Alleviation Control (LAC) strategy, the component dynamic load estimates from the LTI/LQE system are used as feedback signals for limiting or altering control inputs. While LAC offers computationally less intensive solution, it can lead to a conservative design in a specific application at the expense of reduced vehicle maneuver performance. A more effective control strategy for component life extension, albeit at a significant computational complexity, is to limit directly the fatigue life usage associated with harmonic loads that are close in frequency to the dynamic modes of a component. In this Damage Mitigating Control (DMC) strategy, the component dynamic load estimates from the LTI/LQE system are used as inputs to the reduced order structural models for computing estimates of damage variables (for example, D_{comb} described in the section on structural models), which in turn are used as feedback signals for the controller. We will investigate both LAC and DMC strategies in order to assess their performance and computational complexity.

For both LAC and DMC, we will investigate three control schemes, viz., (i) command limiting using a baseline flight control system, (ii) pilot cueing to limit pilot stick displacement and rate, and (iii) closed loop control using Individual Blade control (IBC). We will compare the three control schemes in terms of their performance and impact on vehicle handling qualities. In the command limiting scheme for LAC, we will use inverse of the pre-computed sensitivities between maneuver rates and component dynamic loads in determining maneuver rate limits needed for limiting the dynamic loads, which in turn are used as maneuver command limits for the AFCS. Similarly for DMC, we will arrive at the maneuver command limits using pre-computed sensitivities between maneuver rates and the component damage variable (D_{comb}). In the pilot cueing scheme for both LAC and DMC, the maneuver rate limits are mapped back to pilot stick displacement and rate limits using the *dynamic trim* method. For synthesis of closed loop IBC, we will build on experience from our recent study[6] on optimization of closed loop vibration control with IBC using LTI models and CONDUIT. We will synthesize control laws for harmonic IBC inputs using component harmonic load estimates for LAC and damage variable for DMC as feedback signals.

RESULTS

LTI/LQE Evaluation:

The proposed LTI/LQE scheme was evaluated in simulation for the case of on-line estimation of blade pitch link loads arising from vehicle maneuvers using a nonlinear model of a generic helicopter in FLIGHTLAB®. Using the scheme for LTI model extraction from a nonlinear model presented in Ref. 6, an LTI model with harmonic decomposition of LTP states in a first order representation (i.e., separate displacement and velocity states) was developed from a full vehicle nonlinear (NL) FLIGHTLAB® model of a generic helicopter with elastic blade mode shapes and a 33-state Peters-He dynamic inflow model. The extracted LTI model has previously been validated through comparisons with the nonlinear model both in time and frequency domains and is found to be of sufficient fidelity[6]. The LTP model, developed from the NL model prior to being decomposed into harmonic states for the LTI system, included 8 body states, 33 inflow states (Peters-He Finite state inflow with 4 harmonics and a maximum radial variation power of 8), and 48 multi-blade coordinate (MBC) rotor states including elastic modes. Thus, the total number of LTP states is 89. Each of these LTP states was then decomposed into 0-8/rev harmonic components, resulting in 1513 total LTI model states. The nonlinear model was trimmed for a forward speed of 120 knots. Figure 4 is a plot of the reference blade pitch link axial load variation with time at the selected equilibrium from the nonlinear model.

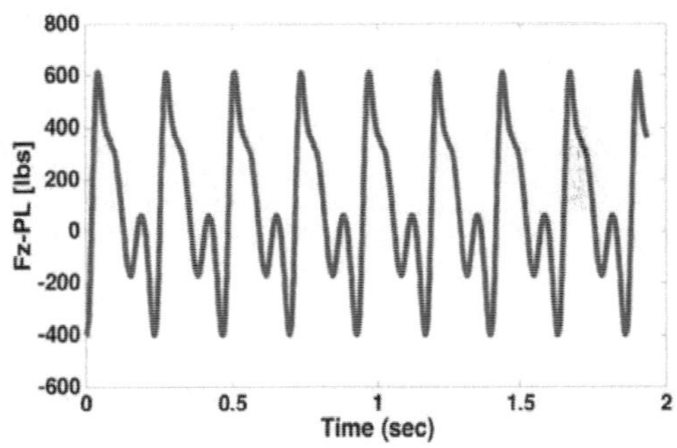

Figure 4. Steady state reference blade pitch link axial load variation with time.

For evaluating the performance of the proposed LTI/LQE scheme, a ramp doublet in longitudinal cyclic control from trim as shown in Fig. 5 was used as input to the nonlinear model as well as the LTI model. The LTI model predictions of the individual harmonic components of fixed system hub shears and moments due to the selected control input were compared with similar predictions from the nonlinear model. The errors in the LTI model predictions of hub shears and moments were used with the Kalman filter (see Figure 3) to correct the LTI model states. The corrected states in turn were used to estimate the reference blade pitch link loads.

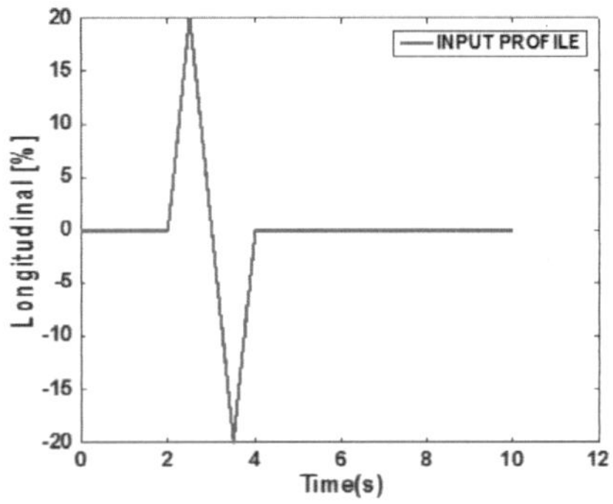

Figure 5. Percentage change from trim of longitudinal cyclic control used for evaluation of the proposed LTI/LQE scheme.

Figure 6 compares the changes in the reference blade 4/rev pitch link loads due to the selected input as predicted by the nonlinear model and the LTI model along with the LTI/LQE estimates. It is seen from Fig. 6 that the proposed LTI/LQE scheme is able to correct the errors in the LTI model predictions of the 4/rev pitch link load predictions. Currently, work is in progress for reducing the size of the LTI models using balanced realization schemes and for evaluating the estimation accuracy through comparison with available pitch link loads test data.

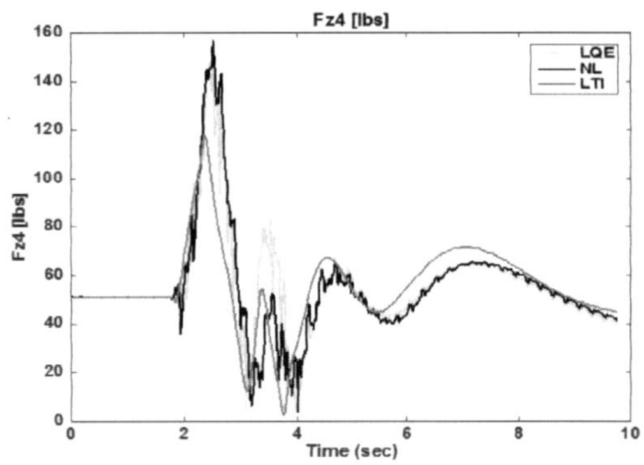

Figure 6. Comparison of reference blade root 4/rev pitch link load variations from trim between nonlinear (NL), LTI and LTI/LQE.

Finite Element Analysis:

In order to establish the need for dynamic finite element analysis for component stresses and strains, a finite element model of a representative geometry of blade root pitch link was developed. The pitch link loads of Figure 4 were used as inputs to the finite element model. The axial load was applied as a concentrated load at the center of the hole of the pitch link. Also the body force associated with centrifugal forces on the pitch link were applied as distributed forces. In the finite element model, only axial displacement and bending rotation (due to centrifugal forces) of the pitch link were included.

Figure 7 presents a comparison of computed stresses at the mid-point location of the pitch link between quasi-static analysis (treating loads to be constant at each instant in time) and dynamic analysis. It is seen from these results that a dynamic analysis is needed for accurately capturing the effects of dynamic loads on resulting stresses and strains. Currently, work is in progress in developing reduced order models based on geometric multi-scale finite element methods.

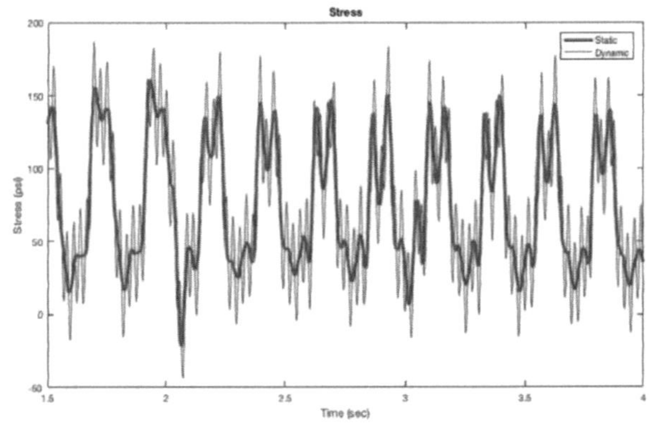

Figure 7. Comparison of stresses at the mid-point of the pitch link from a quasi-static versus dynamic analyses.

CONCLUDING REMARKS

A methodology is presented towards the development of real time algorithms for fatigue critical component life extension in a rotorcraft. The methodology exploits recent developments in reduced order modeling of coupled rotor-body dynamics through harmonic decomposition methods, estimation of rotating component dynamic loads using fixed-system measurements, estimation of component stresses and strains through geometric multi-scale finite element methods, and control schemes for component load alleviation and damage mitigation.

Using simulated data from a nonlinear model of a generic helicopter in FLIGHTLAB®, the proposed LTI/LQE scheme is seen to provide improved estimates of blade root pitch link dynamic load variation to control inputs. Also, from a comparative study of quasi-static versus dynamic analyses, it is seen that a dynamic analysis is needed for accurately capturing the effects of dynamic loads on resulting pitch link component stresses and strains.

Author contact:

JVR Prasad, jvr.prasad@aerospace.gatech.edu

Julian J. Rimoli rimoli@gatech.edu

Massimo Ruzzene, ruzzene@gatech.edu

ACKNOWLEDGMENTS

This study is supported under the NRTC Vertical Lift Rotorcraft Center of Excellence (VLRCOE) from the U.S. Army Aviation and Missile Research, Development and Engineering Center (AMRDEC) under Technology Investment Agreement W911W6-17-2-0002, entitled Georgia Tech Vertical Lift Research Center of Excellence (GT-VLRCOE) with Dr. Mahendra Bhagwat as the Program Manager. The authors would like to acknowledge that this research and development was accomplished with the support and guidance of the NRTC. The views and conclusions contained in this document are those of the authors and should not be interpreted as representing the official policies, either expressed or implied, of the AMRDEC or the U.S. Government. The U.S. Government is authorized to reproduce and distribute reprints for Government purposes notwithstanding any copyright notation thereon.

REFERENCES

[1]D.P. Davies, S.L. Jenkins, F.R. Belben, "Survey of fatigue failures in helicopter components and some lessons learnt," Materials Technology Laboratory, Augusta Westland Limited, UK.

[2]Kaye, M., "Dynamic Health and Usage Monitoring System - Program Update," 15th ERF, 1989.

[3]D.B. Caudle, "Damage Mitigation for Rotorcraft through Load Alleviating Control," M.S. Thesis, The Pennsylvania State University, December 2014.

[4] Lopez, M. J. S. and Prasad, J. V. R., "Linear Time Invariant Approximations of Linear Time Periodic Systems," Journal of the American Helicopter Society, Vol. 62, No. 1, Jan. 2017.

[5]F. Casadei, J. J. Rimoli, and M. Ruzzene, "A geometric multi-scale finite element method for the dynamic analysis of heterogeneous solids," *Computational Methods in App. Mech. and Engg*, 2012.

[6]M. Lopez, J.V.R. Prasad, M.B. Tischler, M.D. Takahashi, and K.K. Cheung, "Simulating HHC/AFCS Interaction and Optimized Controllers using Piloted Maneuvers," AHS Forum, 2015.

Real-time ballistic damage assessment for Rotorcraft

Susheel Yadav, Taru Singhal, Jeffrey Bergman, Grant Chang, Irene Li, Amrita Kumar
Acellent Technologies Inc, Sunnyvale, CA

ABSTRACT

One of the most critical assessments for airworthiness during operation is determining the health condition of the vehicle in real time in order to enable swift and accurate decision making by the pilot or ground operators. The integrity of a rotorcraft can be affected by damage resulting from ballistic impact which can become a safety concern. Therefore, the timely and accurate detection and monitoring of ballistic impact damage is essential. Structural health monitoring offers the promise of a paradigm shift from schedule-driven maintenance to condition-based maintenance of assets. This paper provides an overview of a single combined passive-active system capable of in-situ ballistic impact and damage detection along with trajectory estimation for rotorcraft. This novel system integrates impact detection with active damage scanning and trajectory estimation to inform the end user when an impact has occurred, whether damage has resulted, the damage size and trajectory of the impact through the object.

INTRODUCTION

One of the most critical assessments for airworthiness during operation is determining the health condition of the vehicle in real time in order to enable swift and accurate decision making by the pilot or ground operators. The integrity of a rotorcraft can be affected by damage resulting from ballistic impact which if left unchecked, can become a safety concern for the military personnel and lead to irreversible damage or loss of the platform. Therefore, the timely and accurate detection, characterization and monitoring of ballistic impact damage is essential. Traditional NDI of ballistic impact on the vehicle structure is difficult, if not impossible, to perform in an operational environment due to the inaccessibility of the structure and the difficulty in detecting the damage. Structural Health Monitoring (SHM) using a built-in sensor network offers a promising viable solution to these inspection challenges. [1]

Structural health monitoring (SHM) offers the promise of a paradigm shift from schedule-driven maintenance to condition-based maintenance (CBM) of assets. Built-in sensor networks on the structure can provide crucial information regarding the condition, damage state and/or service environment of the structure. Diagnostic information from sensor data can be used for prognosis of the health of the structure and facilitate informed decision processes with respect to swift decisions for rotorcraft usage. Asset management can be performed based on the actual health and performance of the structure, thereby enhancing safety, minimizing failures and maintenance costs, while maximizing reliability and readiness.

This paper provides an overview of the Damage Assessment and Real-Time Trajectory System (DARTTS) that combines a passive-active system capable of in-situ ballistic impact and damage detection along with trajectory estimation for rotorcraft. This novel system combines an impact detection and localization system with a trajectory estimation program and active damage detection system. The combination of the impact detection and trajectory estimation provides the flight crew with immediate information on the location and probable path of the object through the vehicle while the active damage detection system can inform the team whether or not the object has likely caused a penetration.

As shown in figure 1, on-line trajectory estimation of the ballistic impact is highly desired for aircraft safety in battle situations for direct hit guidance, and early warning systems. Modern shot detection and tracking systems based on audio and infrared (IR) systems are capable of detecting the origin of the shot but lack the ability to determine if and where a shot has impacted the targeted vehicle. By pairing the output of a shot detection systems with an onboard impact detection system the situational awareness of the flight crew can be increased in multiple ways. First, the system can determine if any of the shots are impacting the vehicle. Second, when an impact occurs the crew can be made aware of which portion of the vehicle has been impacted. Thirdly, by combining together with output from the shot detection system a trajectory can be calculated and projected to determine if any vital internal components may have been damaged by the bullet

---

Presented at the AHS Airworthiness and HUMS Technical Meeting, Huntsville, Alabama, USA, February 21-22, 2018.

passing through. Finally, the active damage detection system can detect whether or not the impact has caused a penetration meaning the system will not falsely classify a low velocity strike that does little to no damage as a possibly catastrophic event.

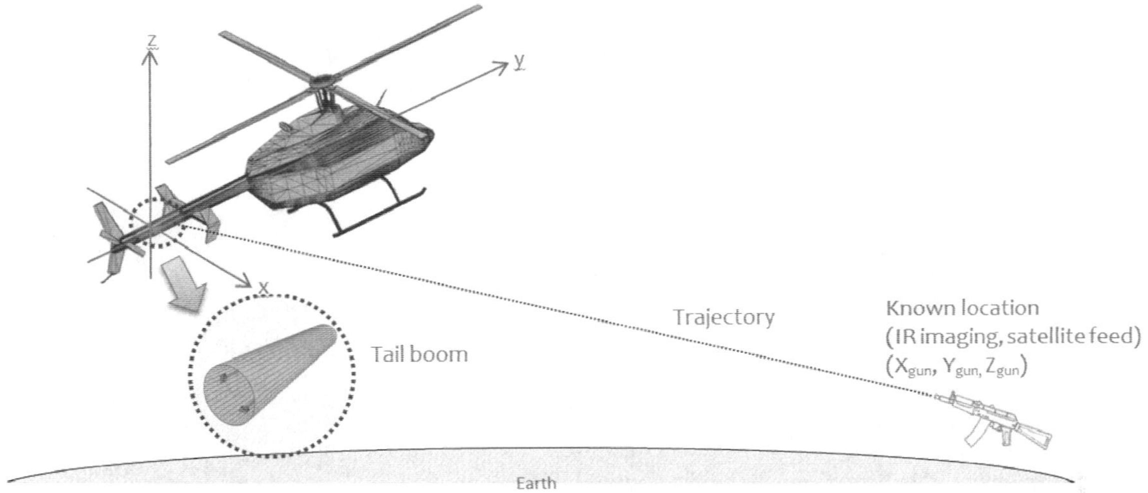

Figure 1: Schematic of ballistic trajectory estimation: a gunshot targeted tail boom section

DARTTS system

The DARTTS SHM system consists of the following 3 components shown in figure 2:

1) *SMART Layer sensors* – The SMART Layer® is well established in the field of Structural Health Monitoring and is currently known for its unique ability to provide a large structural coverage for gathering data with its network of sensors/actuators embedded on a layer thus eliminating the need for each sensor to be installed individually. The layer consists of a network of embedded, distributed piezoelectric discs (PZT) acting as both sensors and actuators for monitoring structural condition in real time.
2) *Lightweight onboard diagnostic hardware* - the passive-active diagnostic hardware is able to monitor structural integrity via passive (impact detection) and active (damage detection) modes.
3) *Impact and damage detection with trajectory estimation software* – the software has features for both active damage detection and passive impact detection, and self-diagnostics. It also features an adaptive algorithm to automatically compensate for damaged sensors and provide reliable damage detection under different environmental conditions.

The passive-active DARTTS system works with support from 3 key capabilities: (1) passive system for impact detection, (2) active system for damage detection, location and sizing, and (3) trajectory estimation enabled by connecting the known gun point location to the impact location and mapping the 2d impact location to the 3d structural component. Customized sensor layers with a *network* of actuators and sensors are integrated with the aircraft structure and combine with the structure to enable the passive impact detection and active damage detection capabilities as explained below.

PASSIVE IMPACT DETECTION: Certain impacts may affect a structure's reliability and integrity over the course of its operational lifetime. It is of utmost importance to be aware of impact events, especially when they happen to rotorcraft in operation. For example, military rotorcraft pose a huge risk of getting hit by ballistics and suffer severe damage. Such damages are nearly impossible to locate during operation and may lead to catastrophic failure of the rotorcraft. Unaddressed hidden damages can result in injuries and fatalities. With information about the location of the ballistic impact, life-saving decisions can be made quicker and also aid in damage control.

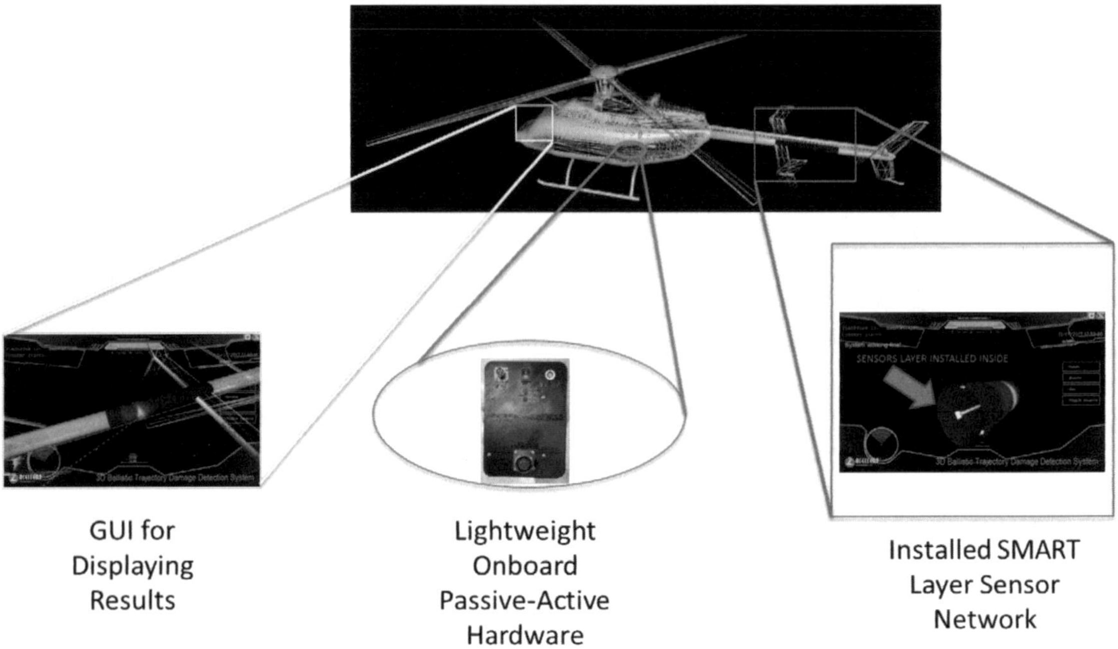

(a) DARTTS Components

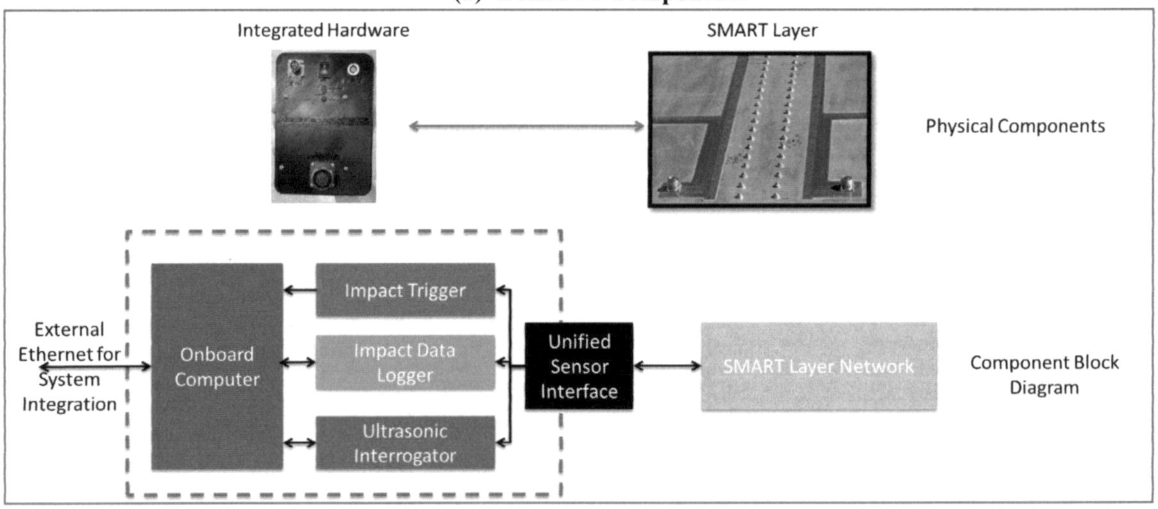

(b) Integration of system physical components consisting of active-passive hardware and SMART Layer sensor networks

Figure 2: DARTTS system

Acellent's passive Impact Monitoring System is set up to be continuously 'listening' for impacts at all times. It is based on impact detection theory with guided lamb-waves. When the structure, is struck by a sufficiently large force by an external mass, the piezoelectric sensors that are bonded onto the structure, pick up the stress waves traveling through the structure. The piezoelectric sensors trigger the system if the voltage generate by the impact is greater than the set threshold level. In other words, impacts that fall below the safety threshold level are ignored, while those that fall above it are reported and generate an alarm signaling the need for the structure's evaluation and repair. After the system is triggered, the data acquisition system collects and stores the data. The data processing algorithm focuses on determining the location of the impact and the impact force time history. The user will be notified of the impact location in real-time.

ACTIVE DAMAGE DETECTION: An active damage detection mode can be used to ultrasonically scan the structure either periodically or immediately after the occurrence of an impact. The results of this scan can provide three levels of damage analysis on the structure.

- Level I: damage detection
- Level II: damage location identification
- Level III: damage size quantification

The functioning of the SMART Layer active system is analogous to that of a built-in *acousto-ultrasonic* system with a network of miniaturized piezoelectric sensors. External factors causing changes in mechanical properties of a structure induce the embedded piezoelectric actuators of the SMART Layer to emit diagnostic signals to neighboring piezoelectric sensors. These signals are subsequently transmitted to the diagnostic hardware for comparison with previously recorded baseline test signals obtained from the structure prior to its application. The diagnostic software uses these signals to identify the location and extent of damage or other structural anomalies. By using different signal processing techniques to process the diagnostic signals, information concerning the structure can be obtained. Customized software algorithms developed to perform these tasks using the diagnostic signals can be used to monitor in-service structures quickly and efficiently. Acellent's active sensing technology is versatile and has been tested for use for damage detection in a variety of structures both metal and composite.

Following each impact event, data is collected from all the actuator-sensor paths in the SMART Layer and compared to the baseline set of data. The active software then analyses the signals to provide information on damage location and size. Figure 3 shows an example diagnostic image after a ballistic impact on a mockup of a tailboom structure.

Figure 3: Example of active damage detection on tailboom structure mapped to 3D structure

TRAJECTORY ESTIMATION: The DARTTS system has an innovative trajectory estimation capability built-in for ballistic impacts. The point B below is the gunshot source location which is known a priori from an outside resource such as IR scanning. The location of point A is determined by the passive system impact detection system. The passive system detects the impact location in a 2D coordinate system. These 2D coordinates are converted to 3D coordinates using surface unwrapping, a 2D→3D mapping scheme, and displayed on the DARTTS software. The trajectory is then projected as a straight line between points A and B and plotted on the DARTTS display

Once the underlying trajectory has been estimated these results can be used to predict the path of the object through the structure allowing the system to predict if any components placed behind the structure are likely to have been impacted.

DARTTS system functioning

An overview of the operational flow for DARTTS can be seen in Figure 4.

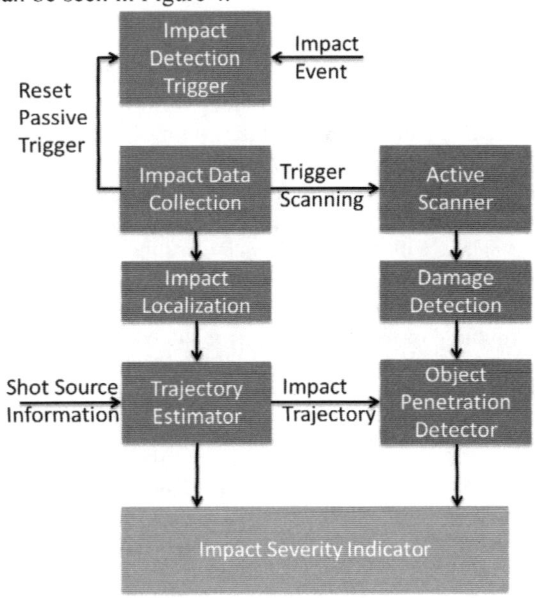

Figure 4: Operational flow for DARTTS

The passive impact detection mode can be used to trigger the active damage detection mode. If the impact recorded is greater than a set threshold value, the impact detection mode will automatically trigger the damage detection mode to enable scanning of the area of impact to determine the extent and size of damage. The active system scans only the structure around the impact and actively determines damage location and extent. This method ensures almost instantaneous damage detection since the system scans only a small region for damage and not the entire structure. The results of this scan can be used to determine if damage has occurred to the monitored structure and if damage has occurred, and whether the ballistic object is likely to have penetrated the structure.

The impact detection and damage detection capabilities are enabled through the pairing of PZT sensors embedded in the SMART Layer sensor networks with both a passive impact data logger and ultrasonic scanning system combined in the same hardware set. An overview of the physical system components which enable this capability can be seen in Figure 2.

The final capability of the system is the trajectory estimation which takes output from the passive and active scanning systems along with the output of a shot origin detector. This portion of the system determines the location of the ballistic impact as determined by the passive system impact detection mode. The passive system detects the impact location in 2D coordinate system. These 2D coordinates are converted to 3D coordinates to show the trajectory of the bullet. Based on the trajectory of the bullet the system can determine the likely path of the projectile through the structure and estimate whether any vital components placed behind the covering structure are likely to have been impacted. These results can then be paired with the active system results to determine the likelihood that the impact has penetrated the structure. Complete information on the location, likelihood of penetration and probable damage due to the impact can the be provided directly to the flight crew who can take remedial actions as necessary.

EXAMPLES

To demonstrate the feasibility of the system a demonstration was conducted on a mockup of a helicopter tailboom consisting of a thin steel duct with SMART Layer sensor networks installed. The SMART Layer sensor network was then connected to the lightweight passive-active hardware for data collection. The passive and active data was passed from the hardware via Ethernet to a laptop where the system analyzed it to determine the location of the impact and whether any damage had occurred. The results of this analysis were then fed into the trajectory estimation algorithm along with an estimated impact origin point to determine the likely trajectory of the object. The final results were then mapped to a three dimensional object model on the GUI and displayed to the end user. As a shot origin detector was not available to Acellent the origin of the ballistic object was taken as an assumption for the purposes of this test. A diagram showing the system components can be seen in Figure 6. For this test two key capabilities were tested:

1. The ability of the system to detect impacts, detect the resulting damage and estimate the trajectory of the object.
2. The ability of the system to detect and localize repeated impacts to the underlying structure while simultaneously actively scanning the structure for damage.

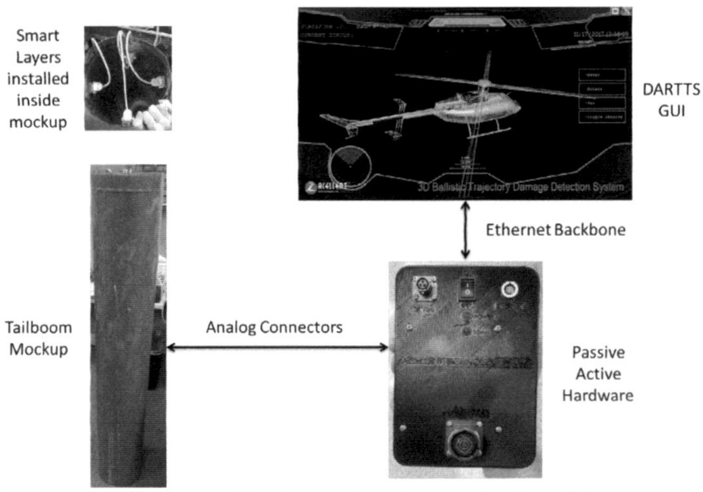

Figure 6: Diagram of test setup for demonstrating DARTTS

Due to the difficulty of using real firearms for testing the system was tested using a high velocity paintball rifle to impact the structure and generate a small non-penetrative damage in the structure. For the test the location of the shooter with the rifle was assumed to be underneath the tailboom structure. During the test the system correctly detected and localized the impact. The system then triggered an ultrasonic scan in response to inspect the damaged region. Based on the ultrasonic scan the system was able to determine that minor damage had occurred to the structure which was displayed on the three dimensional GUI as a low intensity hotspot on a color intensity plot. Furthermore the system showed that the damage and impact locations were coincident showing that the damage detected was caused by the impact.

The output of the GUI can be seen in Figure 7. Overall, the system was found to be able to detect impacts in real-time along with impact location and trajectory estimate results. Active scanning damage results were then quickly and accurately provided to the end user to accurately display the extent of the damage.

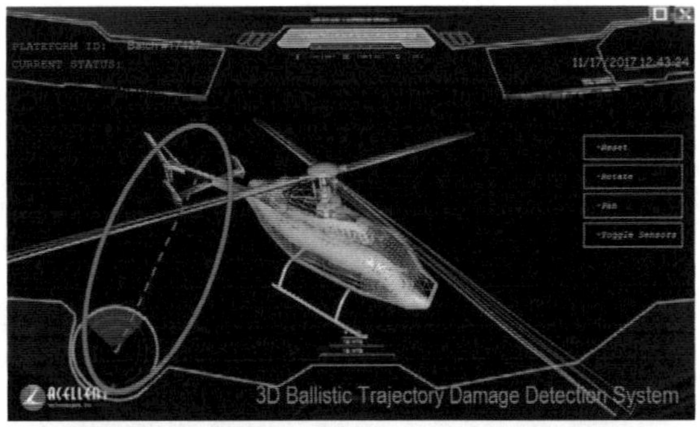

(a) Display of impact detection and trajectory estimation results from test on 3D GUI (note the red circle surrounding the red line denoting the trajectory)

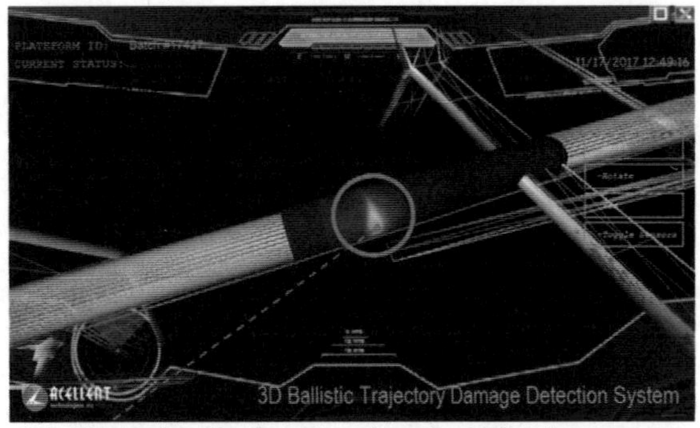

(b) Zoom in display showing the light damage on the structure detected by the active system due to the impact shown as yellow/green discoloration (circled in red) Note the trajectory passes directly through the damaged region.

Figure 7: Results of DARTTS testing with paintball rifle.

CONCLUSIONS

This paper presents a DARTTS SHM system that utilizes three key detection capabilities: (1) a passive system for impact detection, (2) an active system for damage detection, location and sizing, and (3) a trajectory estimation by connecting the known gun point location to the impact location and mapping the 2d impact location to the 3d structure component. Customized sensor layers with a *network* of actuators and sensors were integrated with the aircraft structure at strategic locations in order to enable these 3 detection modes. The system was successfully tested for functionality on a mockup of a rotorcraft tailboom structure.

ACKNOWLEDGEMENTS

The authors would like to acknowledge Xinyi Chen, Pei - Tzu Chen, Matt Kolm, Jordan Yen, Olivia He, Thomas Chen for support on the development.

REFERENCES

1. Structural health monitoring and the detection of ballistic damage, Andrew Baines, Nathaniel Baines, AHS 2011
2. Fatigue Crack Characterization for Rotorcraft Structures under Varying Loads and Temperatures, IWSHM, Sep, 2013.
3. Multi-Sensing Smart Layer Technology for Vehicle Protecting Systems, IWSHM, Sep, 2013.
4. Flight Demonstration of a SHM System on an OH-58 Aircraft, IWSHM, Sep, 2013.
5. An SHM System for detecting Damage in Aging Aircraft, IWSHM, Sep, 2013.

6. Sang Jun Lee, Patrick Pollock, Vishnu Janapati, Franklin Li, Howard Chung, Amrita Kumar, Irene Li, Fatigue Crack Characterization for Rotorcraft Structures under Varying Loads and Temperatures, AHS 69th Annual Forum & Technology Display, May, 2013
7. A. Brookhart and P. Bates, and N. Bordick, "Airframe structural integrity management: A uh-60 application and demonstration," 70th AHS conference, Montreal, Quebec, May 20-22, 2014.
8. Kumar, A., Zhang, D., Li, F., Beard, S., Chung, H., Pollock, P., Lee, S. J., Ikegami, R., Bordick, N., Hodges, C., Jodon, J., "OH-58D Tail Boom Crack Detection System", Proceedings of the AHS Redstone Chapter Technical Specialists' Meeting on Airworthiness, Condition Based Maintenance (CBM), and Health and Usage Monitoring (HUMS), February 11-13, 2013
9. ADS-79-HDBK CBM Aeronautical Design Standard Handbook For Condition Based Maintenance System For Army Aircraft

Regime Recognition Accuracy

Jason Hull & Jeffrey Monaco
Spire Innovations, LLC
Charlottesville, VA, USA

Mark Davis & Raymond Beale
Health Management Systems
Sikorsky Aircraft Corp
Stratford, CT, USA

Roberto Semidey &
Mark Glucksman-Glaser
Structures Division
Naval Air Systems Command
Patuxent River, MD, USA

ABSTRACT

Accurate characterization of fleet and vehicle usage spectrums would allow component retirement times to be based on actual aircraft usage rather than on an assumed worst case usage spectrum used in a traditional time-based maintenance approach. A key enabling technology for such a Usage or Condition Based Maintenance (UBM/CBM) program is Regime Recognition (RR). In general, production RR algorithms with the required accuracy have not been fielded, in part because of the challenges associated with Verification and Validation (V&V). The multi-class classifier analysis methods and associated visualizations presented herein address the need to maximize the utility of available data and rigorously perform V&V of RR algorithms and software. The resultant methods clearly show the performance range that can be expected when an RR code is deployed to the fleet, and provide a single accuracy metric that highlights error characteristics that are most important to a UBM/CBM program.

NOTATION

a	Regime classification accuracy
a_c	Error cost weighted regime accuracy
A	Classifier accuracy
A_c	Error cost weighted classifier accuracy
$C_{i,j}$	Element of the confusion matrix
C_{FN}	Cost of a false negative error
C_{FP}	Cost of a false positive classification error
N	Number of regimes
FN	False negative
FP	False positive
TN	True negative
TP	True positive
α	Regime specificity
$\bar{\alpha}_{micro}$	Classifier specificity from micro average
β	Regime recall rate
$\bar{\beta}_{micro}$	Classifier recall rate from micro average
ρ	Regime operating point
$\bar{\rho}$	Classifier operating point
ρ_c	Error cost weighted regime operating point
$\bar{\rho}_c$	Error cost weighted classifier operating point

INTRODUCTION

Regime Recognition (RR) is a key enabling technology for a Usage or Condition Based Maintenance (UBM/CBM) program. Compared to traditional time based maintenance, which derives component retirement times from an *expected* usage spectrum, the goal of a UBM/CBM approach is to calculate the usage and/or loads profile(s) of individual vehicles and its subsystems based on the actual operating conditions (i.e., regimes) and/or loads experienced by the aircraft.

A challenge to effective UBM is accurately capturing the usage or load profiles. To date, it has been impractical to instrument fleet aircraft to directly measure all the loads that would be required to monitor key life-limited components. However, the ability to collect aircraft state data that can be used for usage monitoring has become practical with the integration of state-of-the-art Health and Usage Monitoring Systems (HUMS) on a number of aircraft models. Many of the deployed HUMS have generic algorithms that can classify the regimes encountered in flight from the available aircraft sensors. If the resultant usage spectrum is accurate, then component life expenditure can be calculated as a function of the duration and number of occurrences of each flight regime and the strains/loads associated with each flight regime (as determined using flight loads survey data collected during aircraft development and certification). The challenge is then to derive algorithms that accurately infer the regime profile, ideally in terms of both occurrences (regime counts) and duration (time spent in the regime).

Presented at the AHS Airworthiness and HUMS Technical Meeting, Huntsville, AL, Feb 21-22, 2018.

Thus, in lieu of direct load measurements, a UBM program requires an RR algorithm that can accurately characterize the regimes for a particular vehicle. In general, production algorithms with the required accuracy have not been fielded, in part because of the expense and human involvement currently required to develop and independently verify and validate the candidate approaches. A dominant factor is the cost of flight tests and the associated time-consuming V&V process to manually compare RR algorithm results with flight test data. Development and validation of RR algorithms can leverage flight load survey (FLS) tests that are conducted to achieve aircraft qualification/certification. Such data, however, are potentially deficient in three respects, (1) the tests focus on critical regimes and corner points of the operational envelope and may not closely represent the region of the operational envelope where fleet aircraft tend to operate, (2) there may be insufficient maneuvers flown by multiple pilots to rigorously perform RR V&V with sufficient variation for all maneuvers, and (3) maneuvers are flown very precisely and perhaps somewhat differently from how fleet aircraft are flown. Two distinct yet complimentary approaches have been developed by the authors to address these deficiencies. The first approach is to augment the FLS evaluation database with independent maneuver examples from model-based rotorcraft simulations. This capability is then complimented with analysis methods and automated processes that maximize the utility of available data to support the rigorous V&V of RR algorithms and software. The focus of this paper is on the analysis methods.

The first step in the V&V process is to reconcile the regime labels produced by the classifier and the FLS regimes, which correspond to the design usage spectrum used to life aircraft components. The objective here is to establish a mapping between these potentially disparate labels to facilitate a direct comparison between the output of the classifier with the truth regimes in the FLS evaluation database and the regimes associated with component damage in the design usage spectrum. The need for this is critical if the RR algorithms were developed by a HUMS supplier for generic application to any aircraft model, which is typical today. Ideally, an FLS evaluation database includes maneuvers executed at different altitude, and weather conditions and by several pilots at a variety of loadout conditions to affect vehicle weight and center of gravity location.

Once a common set of regime labels is established, the FLS regimes in the evaluation database can be compared to the classifier output to produce a confusion matrix. From this matrix, more than a dozen statistics can be derived that quantify different aspects of classifier performance. The question then becomes which metric or metrics are important for regime classifiers. This question is further complicated by the fact that we are dealing with multi-class classifiers that will likely include more than 100 unique regimes. The Army's Aeronautical Design Standard ADS-79-HDBK Handbook (Ref. [1]) provides guidance on this question with a stated requirement to develop a V&V process that promotes RR algorithms with a minimum accuracy of 97% and that err on the side of conservatism by selecting severe (damaging) regimes.

In this paper we present analysis methods that address this question given the requirements stated in Ref. [1]. These methods leverage the cost curves developed by Drummond & Holte in Ref. [2] for binary classifiers to develop *accuracy curves* for multi-class classifiers. These curves provide a natural visualization framework that shows how the accuracy metric referenced in the ADS-79-HDBK handbook is biased by the prevalence of maneuvers in an FLS evaluation database. This bias is removed via error cost weighting to impart application specific knowledge into the V&V process to obtain a classifier accuracy score that is most relevant to a UBM program. In so doing, this single score satisfies both objectives of the ADS-79-HNBK handbook stated above. The accuracy curve visualizations developed to support a multi-class classifier also shows how an RR algorithm will perform under all possible maneuver distributions encountered when it is deployed to the fleet. This is an important piece of information if one is interested in how well a candidate RR algorithm will estimate the fleet usage spectrum. The visualizations include confidence intervals, which will allow an analysis to compare candidate RR algorithms and determine if there is a statistically significant difference between them.

ACCURACY CURVES

To introduce the concept of accuracy curves we employ a simple, illustrative example in which a multi-class regime recognition algorithm, RR1, is tasked with the identification of high damage regimes, mild damage regimes, and benign regimes. The classification results for this conceptual example are presented in the confusion matrix of Table 1. The confusion matrix, also referred to as the contingency table or coincidence matrix, is commonly used in the fields of machine learning and data mining to visualize and quantify the performance of a classification algorithm. The rows show the actual regime occurrences, i.e. truth; the columns show the regime occurrences identified by the RR code. The values represent time in minutes for this example. For transient regimes, these values could also represent the number of occurrences.

Table 1: Confusion Matrix RR1 Classifier

Confusion Matrix		RR Classification		
		Benign	Mild Damage	High Damage
Truth	Benign	380	8	3
	Mild Damage	4	53	4
	High Damage	0	13	20

To illustrate what this matrix conveys, we make the following observations,

- there was a total of 391 minutes of benign maneuvers in the evaluation database, and for 380 of those minutes the RR algorithm correctly classified them as benign and incorrectly classified 8 minutes as mild damage flight time and 3 minutes as high damage flight time;

- there was a total of 61 minutes of mild damage regimes in the evaluation database and 53 of those minutes were correctly classified by the RR algorithm while 4 minutes were incorrectly classified as benign and 4 minutes were incorrectly classified as high damage;

- there was a total of 33 minutes of high damage regimes in the evaluation database and 20 of those minutes were correctly classified by the RR algorithm while 13 minutes were incorrectly classified as mild damage;

- benign regimes made up 81% of the evaluation database;

- mild damage maneuvers made up 12% of the evaluation database;

- high damage maneuvers made up 7% of the evaluation database.

To develop the statistics necessary to quantify the performance of an RR algorithm, we first total all the correct and incorrect classification results for each regime. There are two types of correct classification results:

- True Positive (TP) – the regime is occurring and the RR algorithm correctly reports that it is occurring

- True Negative (TN) – the regime is not occurring and the RR algorithm correctly reports that it is not occurring

and there are two types of classification errors:

- False Positive (FP) – the regime is not occurring and the RR algorithm incorrectly says that it is occurring. This is also referred to as a Type I error.

- False Negative (FN) – the regime is occurring and the RR algorithm incorrectly reports that it is not occurring (i.e., it misses the regime). This is also referred to as a Type II error.

The classification totals for our classifier are given below in Table 2.

Table 2: RR Classification Totals for RR1 Classifier

	TP	TN	FP	FN
Benign	380	90	4	11
Mild Damage	53	403	21	8
High Damage	20	445	7	13

From these totals, we show how to calculate three accepted statistics that capture important performance information for an RR algorithm. For a multi-class classifier, these statistics are calculated for each regime in a one versus all strategy. When performing the calculation for each regime, we refer to the regime under evaluation as the relevant regime. The remaining regimes are referred to as non-relevant regimes. In Figure 1, we organize the relevant and non-relevant regimes with respect to the four possible classification outcomes to highlight the subpopulations that are used to calculate the performance statistics. In this illustration, benign regimes are under evaluation, therefore the subpopulation of relevant regimes (benign) are in the left hemisphere. All non-relevant regimes are on the right. The regimes located in the inner circle are those that were identified by the RR algorithm as benign. Each colored marker represents one minute of flight time.

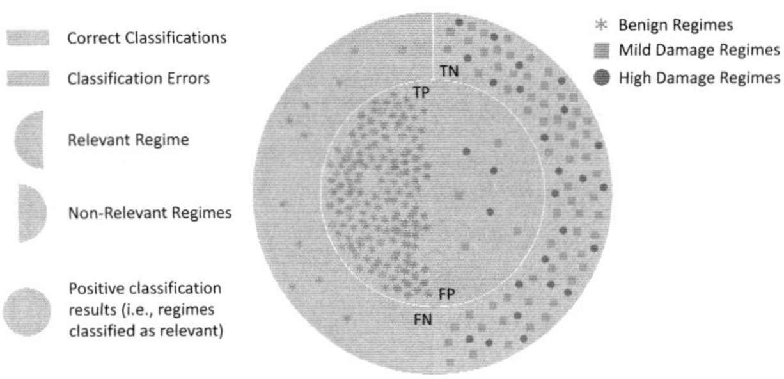

Figure 1: Classification Regions and Sub-populations

Recall

Recall provides an accuracy metric for the population of flight time shown in the left hemisphere of Figure 1 (i.e., the flight time that is spent in the relevant regime). For a given regime, the recall accuracy metric is a measure of the fraction of relevant regimes that were correctly retrieved by the classifier. This is quantified as,

$$\beta = \frac{TP}{TP + FN}. \quad (1)$$

From this expression we can see that recall is an accuracy metric that is driven down by regime misses (i.e., FN errors). Therefore, if it is important to know how likely a candidate RR algorithm will not miss a regime (e.g., because it is highly damaging), this is an important accuracy metric to know.

Specificity

Specificity provides an accuracy metric for the population of flight time shown in the right hemisphere of Figure 1 (i.e., the flight time that is not spent in the relevant regime). For this population, specificity gives the rate of correct classification of negative events (i.e., the proportion of non-relevant regimes that are correctly classified as non-relevant). This is formalized as,

$$\alpha = \frac{TN}{TN + FP}. \quad (2)$$

From this expression it is clear that specificity is an accuracy metric that is driven down by FP errors (i.e., the overestimation of the relevant regime). In the context of a UBM program, an accuracy metric that is dependent on FP errors is particularly useful if it is important that the time spent in a particular regime not be overstated because it would result in the premature replacement of non-safety-critical components. Furthermore, any regime in the right hemisphere of Figure 1 is a potential FP error and not all FP errors are equal. For example, if the high damage regimes that are missed are benign FP errors, more damage would be underestimated than if they were mild damage FP errors.

Accuracy

The overall regime accuracy employs the entire population of regimes shown in Figure 1. It is defined as the proportion of all the classification results (TP, TN, FP, and FN) that are correct (TP, TN),

$$a = \frac{TP + TN}{TP + TN + FP + FN}. \quad (3)$$

This overall accuracy metric can be expressed as a function of recall and specificity. To obtain this expression we first define a regime operating point, ρ, as the percentage of flight time spent in the relevant regime (i.e., the prevalence of the regime). From Figure 1 it can be expressed as,

$$\rho = \frac{TP + FN}{TP + TN + FP + FN}. \quad (4)$$

With this definition, the overall accuracy for the relevant regime can be expressed as follows,

$$a = \beta\rho + \alpha(1 - \rho). \quad (5)$$

Thus as the operating point varies over the domain $\rho = [0,1]$, an accuracy curve defining the performance of the classifier is obtained.

The accuracy curves for all three regimes are shown in Figure 2 and the statistics are listed in Table 3. For each curve, we show the operating point, (ρ, a), with a marker. This figure also includes accuracy curves for the trivial "Always True" and "Always False" classifiers. This is an important concept, and we can see that under some operating conditions (evaluation database configurations), i.e. operating points to the left and right of the always false and always true classifiers, respectively, the trivial classifiers will have a higher accuracy than any candidate RR algorithm. The region below these trivial classifiers is defined as,

$$\beta = 1 - \alpha, \quad (6)$$

which is to say this is the *chance* region where the classifier is equally as likely to produce a true positive as false positive. If the operating accuracy is in this region then the classifier is no better than a chance guess.

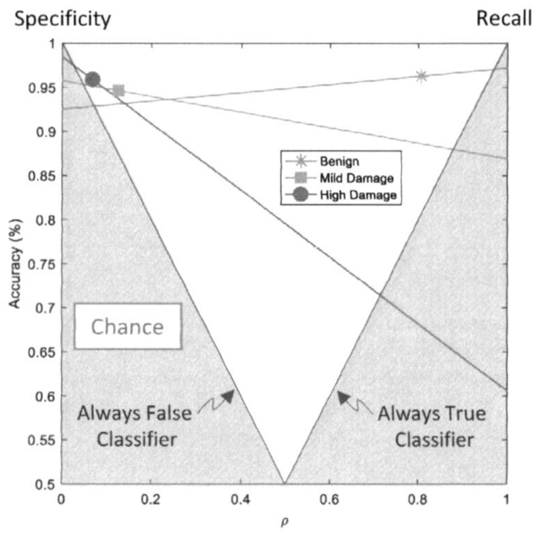

Figure 2: Regime Accuracy Curves for RR1 Classifier

The horizontal axis in Figure 2 can be interpreted as the rate at which a regime occurs on an individual aircraft, or in the fleet. The horizontal location of the operating points on each accuracy curve mark the prevalence of the corresponding regimes in the FLS evaluation database used to assess the regime recognition algorithm. As such, the accuracy curves start to give us a sense of how regime accuracy is influenced by the regime prevalence.

The horizontal axis can also be interpreted as an indication of the relative importance of recall versus specificity. If the regime does not appear often, then the operating point will be close to zero. As such Eq. (5) will be weighted to favor specificity, and the resulting regime accuracy will be a measure of how likely a regime will not be over-counted via FP errors. Intuitively this makes sense; if we were concerned only with such FP errors for a particular regime, we would use an evaluation database that had no occurrence of the regime. In so doing the only possible errors would be FP errors and Eq. (5) would collapse to be a measure of specificity. It should be noted that if the importance of avoiding FP errors is extreme enough to warrant an operating point of zero then Figure 2 implies that a trivial Always False classifier should be employed.

Conversely, if a regime is heavily represented in the database, the operating point will be close to one. As such Eq. (5) will be weighted to favor recall and the resulting accuracy will be a measure of how likely a regime will be retrieved by the classifier. Again, intuitively this makes sense; if we were concerned only with the probability that a regime will not be missed, we would use an evaluation database that only contained that regime. In so doing, the only possible errors would be FN errors and Eq. (5) would collapse to be a measure of recall. It should again be noted that if the importance of avoiding FN errors is extreme enough to warrant an operating point of unity, then we can see from Figure 2 that a trivial Always True classifier should be employed.

Table 3: Accuracy Statistics

Regime	Accuracy	Specificity	Recall	ρ
Benign	0.97	0.96	0.97	0.81
Mild Damage	0.94	0.95	0.87	0.12
High Damage	0.96	0.98	0.61	0.07

Showing the overall regime accuracy statistic within the context of accuracy curves provides valuable information when considering a candidate RR algorithm for use in a UBM program. From Figure 2 we can make the following important observations to accompany the statistics listed in Table 3:

- the accuracy score for benign regimes favors recall and is therefore more a measure of how likely benign regimes will be accurately detected. Put another way, 97% of the benign regimes are likely to be retrieved by the classifier.
- the accuracy score for mild damage regimes favors specificity and is therefore more a measure of how likely mild damage regimes will not be over-counted. Put another way, an accuracy score of 94% is conveying that 6% of the flight time that is not truly a mild damage regime is likely to be classified as such.
- the accuracy score for high damage regimes also favors specificity and is therefore more a measure of how likely high damage regimes will not be over-counted. Put another way, an accuracy score of 96% is conveying that 4% of the flight time that is not truly high damage will likely be classified as such.

From the visualization provided by accuracy curves, we can see how one could manipulate the distribution of maneuvers in the FLS evaluation database to obtain regime accuracy scores that are dependent on error characteristics that are more important to a UBM program. For example, if it is important that high damage regimes are not missed then one could increase the number of high damage regimes in the database to obtain an accuracy score that is more dependent on recall.

Error Cost

From the observations above we can leverage accuracy curves to provide the information necessary to satisfy the secondary objectives in the ADS-79-HDBK handbook to identify RR algorithms that err on the conservative side (i.e., classification errors result in the overestimation of aircraft damage). In our example, the regime accuracy is nearly sufficient in terms of the 97% performance threshold, but the important secondary objective is not satisfied. We can see from the right end of the accuracy curves that there is a relatively high probability that mild and high damage regimes will be missed. This is quantified in Table 3, which states that only 87% of mild damage regimes will likely be detected, and only 61% of high damage regimes.

Our goal then is to obtain an accuracy metric that leverages the information available in the accuracy curves to capture and quantify this secondary objective. To do so we introduce the concept of the cost of a classification error. This concept will allow us to move each operating point to the region of the accuracy curve that conveys this critical information. To introduce this concept we express the overall accuracy in Eq. (5) in terms of the FN $(1-\beta)$ and FP $(1-\alpha)$ error rates,

$$a_c = 1 - \left[\frac{(1-\beta)\rho C_{FN} + (1-\alpha)(1-\rho)C_{FP}}{\rho C_{FN} + (1-\rho)C_{FP}}\right] \tag{7}$$

where C_{FN} is the cost of an FN error and C_{FP} is the cost of an FP error. Nominally these parameters are both set to one and Eq. (7) collapses to Eq. (5). If FN errors are very costly (i.e., they would lead to the underestimation of aircraft damage) then C_{FN} can be increased relative to C_{FP}. This effectively moves the operating point to the right to obtain an accuracy score that is more dependent on FN errors. This movement is clear if we define a cost weighted operating point as,

$$\rho_c = \frac{\rho C_{FN}}{\rho C_{FN} + (1-\rho)C_{FP}} \tag{8}$$

Then Eq. (7) can be simplified to,

$$a_c = \beta\rho_c + \alpha(1-\rho_c) \tag{9}$$

We can now leverage this concept of the cost of an error to impart general knowledge of the environment that a candidate RR algorithm will be operating in. For our example, this information could include general UBM knowledge such as,

1. it is important not to miss high damage regimes as this would lead to the underestimation of damage for safety critical components;
2. it is important not to miss mild damage regimes since it would also lead to the underestimation of damage; however, it is less important than high damage regimes
3. it is important not to overstate the presence of benign regimes as this may also lead to the underestimation of aircraft damage

To impart the first piece of operational knowledge we increase C_{FN} relative to C_{FP} for high damage regimes to indicate that recall is more important than specificity. To impart the second piece of operational knowledge, we also increase C_{FN} relative to C_{FP} for mild damage regimes. The final piece of operational knowledge is imparted by increasing C_{FP} relative to C_{FN} for benign regimes to indicate that specificity is more important than recall. The results of this error weighting are shown below in Figure 3.

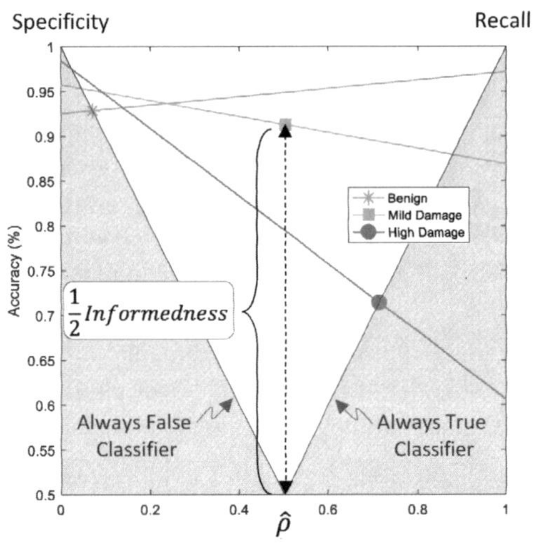

Figure 3: Accuracy Curve with Error Cost Weighting for RR1 Classifier

Increasing the cost of missing a high damage regime moves the operating point to the right to increase the influence of recall on the overall accuracy metric. Since FN errors were deemed to be most critical, C_{FN} is increased to the intersection with the Always True trivial classifier. This is a natural boundary since moving beyond this point would imply that not missing a high damage regime is so important that we should assume the vehicle is always in a high damage regime. The operating point for mild damage regimes is also moved to the right to obtain an accuracy score that is more dependent on recall. Here we chose an error cost weighting that centers the operating point at 0.5 thereby removing the dependence on the prevalence of this maneuver class in the evaluation database. Consequently, this operating point corresponds to the location of the Informedness statistic,

$$Informedness = \beta + \alpha - 1 \tag{10}$$

Thus, in the context of accuracy curves, the Informedness statistic is found by first applying an error cost weighting such that $\hat{\rho} = 0.5$ and then doubling the distance from the point on the accuracy curve to the boundary of the chance region. As such, this statistic quantifies how informed the output of a candidate RR algorithm is versus a chance guess [3].

The operating point for benign regimes moves to the left to increase the influence of FP errors on the regime accuracy score. Here again, we move to the intersection with the Always False trivial classifier since moving beyond this point would imply that it is so important not to over-count benign regimes that we should use a classifier that reports they never occur.

The resulting statistics are listed below in Table 4. Note that specificity and recall have not changed, but the accuracies for each regime have gone down to reflect the fact that the secondary objective of erring on the side of overestimating damage has not been satisfied. We can see that the ability to move around on the accuracy curves provides a means to highlight the error characteristics that are important to a UBM program and that a candidate RR algorithm with superior performance at these relevant operating points would rise to the top.

Table 4: Accuracy Statistics with Error Cost Weighting for RR1 Classifier

Regime	Error Cost Weighted Accuracy	Nominal Accuracy	Specificity	Recall	ρ_c
Benign	0.93	0.97	0.93	0.97	0.07
Mild Damage	0.91	0.94	0.96	0.87	0.50
High Damage	0.71	0.96	0.98	0.61	0.71

It should be noted that similar movements in the operating points could be made by simply changing the prevalence of regimes in the FLS evaluation database to establish which types of errors are most important. Given the limited availability of flight test data due to the expense of collecting and verifying the data, it is likely that the ability to manipulate the distribution of regimes would be limited. Error cost weighting provides a formal approach to achieve the same goal without manipulating the makeup of the database – thus maximizing the utility of the available evaluation dataset. Furthermore, this process allows us to specify error importance configurations that could not be achieved via a manipulation of the FLS database.

Multiclass Accuracy Metrics

In practice, RR algorithms are tasked with the classification of a hundred or more unique regimes. Reviewing and comparing accuracy scores for each regime to determine which RR algorithm is superior would be impractical. Therefore, we seek an overall classifier accuracy score that captures performance and incorporates the error cost weighting to ensure the result is relevant to a UBM program.

When calculating average accuracy metrics for a multi-class classifier there are two methods to consider – micro averaging and macro-averaging (see Ref. [4]). Micro-averaging gives equal weighting to each regime occurrence; macro-averaging gives equal weighting to each regime class. For any evaluation measure, f, calculated as a function of the TP, TN, FP, and FN totals (e.g., specificity, and recall) the macro average is given by the arithmetic mean of the individual regime metrics,

$$\bar{f}_{macro} = \frac{1}{N}\sum_{i=1}^{N} f(TP_i, TN_i, FP_i, FN_i) \tag{11}$$

where N is the number of regimes. The micro average is given by,

$$\bar{f}_{micro} = f\left(\sum_{i=1}^{N} TP_i, \sum_{i=1}^{N} TN_i, \sum_{i=1}^{N} FP_i, \sum_{i=1}^{N} FN_i\right) \tag{12}$$

The accuracy statistic given in Eq. (3) is actually invariant to the type of averaging used to aggregate the individual regime accuracy scores. Whether micro or macro averaging is employed, the overall classifier accuracy, A, can be expressed as,

$$A = \frac{\sum_{i=1}^{N}\sum_{j\neq i} C_{ij} + N\sum_{i=1}^{N} C_{ii}}{N\sum_{i=1}^{N}\sum_{j=1}^{N} C_{ij}} \tag{13}$$

where C_{ij} is the element of the confusion matrix in the ith row and j^{th} column. If we want to express A in terms of the average recall and specificity,

$$A = \bar{\beta}_{micro}\bar{\rho} + \bar{\alpha}_{micro}(1 - \bar{\rho}) \tag{14}$$

where $\bar{\rho} = \frac{1}{N}$ is the average operating point, then $\bar{\alpha}$ and $\bar{\beta}$ must be micro averages. The micro-average for recall can also be expressed as,

$$\bar{\beta}_{micro} = \boldsymbol{\rho}^T\boldsymbol{\beta} \tag{15}$$

where $\boldsymbol{\rho} = [\rho_1, \rho_2, \dots, \rho_N]^T$, and $\boldsymbol{\beta} = [\beta_1, \beta_2, \dots, \beta_N]^T$, which shows that giving equal weight to each regime occurrence effectively amounts to a weighted average where the weights are given by the prevalence of each regime in the evaluation database. It is important to note that micro average for recall can also be expressed as,

$$\bar{\beta}_{micro} = \frac{\sum_{i=1}^{N} C_{ii}}{\sum_{i=1}^{N}\sum_{j=1}^{N} C_{ij}}, \tag{16}$$

which is to say that it is the sum of the diagonal elements of the confusion matrix divided by the sum of all elements of the confusion matrix. This is a common metric used to quantify the performance of a multi-class classifier (See Ref. [5] and [6]). It will be shown, however, that relying on this metric alone and ignoring the specificity metric in the second term of Eq. (14) can mask error characteristics that are important to a UBM program.

The micro-average of specificity can be expressed as,

$$\bar{\alpha}_{micro} = \frac{(\mathbf{1} - \boldsymbol{\rho})^T}{N - 1}\boldsymbol{\alpha} \tag{17}$$

where $\boldsymbol{\alpha} = [\alpha_1, \alpha_2, \dots, \alpha_N]^T$, which also indicates a dependence on the prevalence of regimes in the evaluation database. Using these relationships, the overall accuracy in Eq. (14) can be expressed as,

$$A = \sum_{i=1}^{N} \omega_i\big(\beta_i\rho_i + \alpha_i(1 - \rho_i)\big), \tag{18}$$

where $\omega_i = 1/N$. When the error costs are applied to the underlying regime accuracy curves, we know from Eqns. (8) and (9) that only the operating points move and specificity and recall remain fixed. As such, this cost information is also imparted to the overall classifier accuracy via the cost weighted operating points,

$$A_c = \sum_{i=1}^{N} \omega_i \left(\beta_i \rho_{c_i} + \alpha_i (1 - \rho_{c_i}) \right) \tag{19}$$

The overall cost weighted operating point is given by,

$$\bar{\rho}_c = \sum_{i=1}^{N} \omega_i \rho_{c_i} \tag{20}$$

Nominally, $\omega_i = 1/N$; however, it can also be used to weight the relative importance of each regime class.

This overall classifier accuracy is shown in Figure 4 for both the nominal error cost weighting (i.e., $C_{FN} = C_{FP} = 1$ for all regimes), indicated by the solid blue circle marker, and the error cost weighing depicted in Figure 3 (magenta circle marker). In both cases, $\omega_i = 1/N$. Since the error cost weighted accuracy is to the right of the nominal accuracy we know that it has an increased dependence on recall.

For a given evaluation database there is an infinite number of error cost configurations for the underlying regime accuracy curves with each one resulting in some $(\bar{\rho}_c, A_c)$ pair. The region of achievable $(\bar{\rho}_c, A_c)$ pairs is bounded by a convex hull with $2N$ vertices that span the entire operating range. In Figure 4, the convex hull boundary for our example is shown in black. The overall error cost weighted classifier accuracy can be anywhere in this region. The location is dependent on the distribution of maneuvers in the evaluation database and the error costs assigned to each regime. If the operating point for the accuracy is biased to the left we know the result is driven by FP errors and is more a measure of specificity; if it is biased to the right we know it is driven by FN errors and is more a measure of recall. In either case, we know that it highlights error characteristics important to the application.

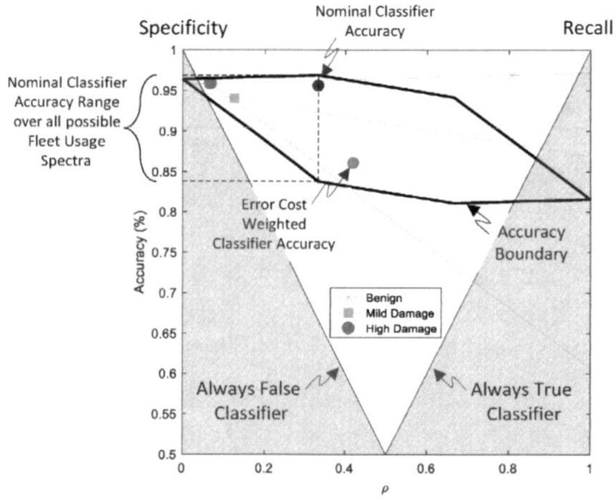

Figure 4: Overall Classifier Accuracy and Cost Weighted Accuracy Boundary for RR1 Classifier

This presentation of the accuracy statistics also allows us to identify the classifier accuracy range over all possible distribution of maneuvers that could be encountered when an RR algorithm is deployed to the fleet. This is given by the range of accuracy scores bounded by the convex hull and at the $\bar{\rho} = \frac{1}{N}$ operating point. From Figure 4 we can see that the deployed usage spectrum can have a significant impact on the accuracy. This is an important piece of information when selecting an RR algorithm to estimate the actual fleet usage spectrum. One would seek an algorithm with sufficient accuracy and with a small accuracy range at this operating point. These classifier accuracy statistics are listed in Table 5.

Table 5: Overall Classifier Accuracy Statistics

Statistic	Value
Minimum deployed accuracy	0.86
Maximum deployed accuracy	0.97
Accuracy at nominal operating point	0.96
Accuracy at UBM operating point	0.85

For classifier RR1 the error cost weighting drove down the accuracy score from 96% to 85% because the error properties important to a UBM program were not as good as the overall error characteristics highlighted by the prevalence of maneuvers in the evaluation database. We can see that this accuracy could also have been found via proper manipulation of the FLS evaluation database as it is within the accuracy range along the dashed line in Figure 4. We can also see that manipulating the database cannot cover the entire accuracy range that is achievable through error cost weighting. Thus, error cost weighting allows us to maximize the utility of the available data by going beyond what is possible via the manipulation of the evaluation database and impart any information that could be relevant to a UBM program.

In this illustrative example, when high damage regimes are missed, they are classified as mild damage regimes. Now suppose we must choose between this classifier and a competing RR algorithm, RR2, that misses the same number of high damage regimes (i.e., has the same recall), but these misses are classified as benign regimes (see confusion matrix in Table 6).

Table 6: Confusion Matrix for RR2

Confusion Matrix		RR2 Classification		
		Benign	Mild Damage	High Damage
Truth	Benign	380	8	3
	Mild Damage	4	53	4
	High Damage	13	0	20

Clearly RR1 should be selected as RR2 would underestimate more damage than RR1. However, if we are only looking at the overall accuracy given by Eqns. (13), (14), or (16), as is commonly done, this important distinction would be missed. Indeed, all three metrics are invariant to such an important change in the confusion matrix. If, however, we apply the error cost weighting in Eq. (19), to impart UBM knowledge we get the accuracy curves shown in Figure 5 and the accuracy statistics listed in Table 7.

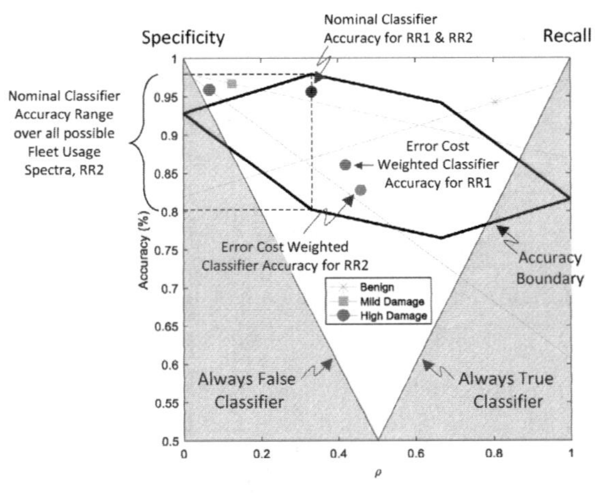

Figure 5: Nominal and Error Cost Weighted Accuracy for RR1 versus RR2

As previously stated, the nominal classifier accuracy remains unchanged at 96% as a result of this change in the confusion matrix. The error cost weighted accuracy for RR2 at the UBM operating point, however, is three percentage points lower than RR1. Furthermore the accuracy range for RR2 is greater than RR1 – an indication that we can expect more variability in the performance of RR2 when deployed to the fleet.

Table 7: Overall Classifier Accuracy Statistics

Statistic	RR1	RR2
Minimum deployed accuracy	0.86	0.80
Maximum deployed accuracy	0.97	0.98
Accuracy at nominal operating point	0.96	0.96
Accuracy at UBM operating point	0.85	0.82

From this illustrative example it is clear that the proposed analysis methods greatly facilitate the comparison of RR algorithms with a single accuracy score. Error cost weighting allows us to focus on the region of the accuracy space most important to a UBM program to highlight even the most subtle differences in error characteristics that are critical to the estimation of aircraft fatigue damage.

EXAMPLE

We now apply these analysis methods to results from a prototype reconfigurable RR tool developed by Sikorsky Aircraft Corporation (SAC). For this sample problem, SAC configured the RR tool to classify 16 of 100 plus regimes contained in the UH-60 usage spectrum as given in Table 8. The configuration reflects an approach similar to that used by HUMS suppliers to provide a generic onboard RR capability that has not been customized or optimized for the UH-60.

Table 8: Regime List

Regime
Climb
Level Flight
Sideslip
Dive
Partial Powered Descent
Autorotation
Climb Left Turn
Climb Right Turn
Left Turn 30°
Left Turn 45°
Left Turn 60°
Right Turn 30°
Right Turn 45°
Right Turn 60°
Pullout
Pushover

The evaluation database used to generate the confusion matrix included 1.25 hours of FLS regimes provided by SAC and was augmented with 1.75 hours of simulated maneuvers. In Figure 6, we show the individual regime accuracy curves along with the convex hull that bounds the classifier accuracy for our evaluation database. The classifier accuracy associated with the error cost implicitly defined by the distribution of maneuvers in the evaluation database is marked as the nominal classifier accuracy.

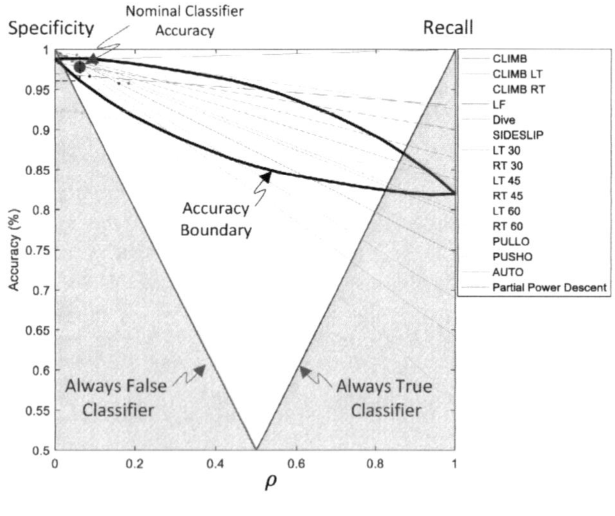

Figure 6: Regime Accuracy Curves and Classifier Accuracy Convex Hull for UH-60 IVHMU Onboard Regime Recognition Algorithm

In

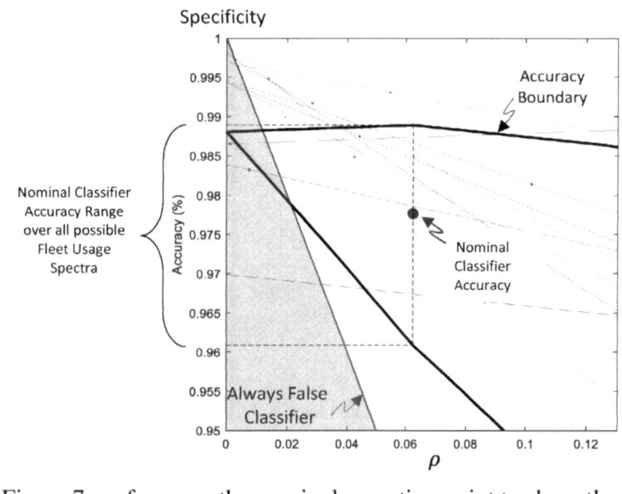

Figure 7 we focus on the nominal operating point to show the classifier accuracy range over all possible regime distributions that could be encountered when deployed to the fleet. As the number of regimes identified by the classifier, N, increases, the nominal operating point, $\bar{\rho}$, will move to the left and the region inside the convex hull will naturally decrease. Thus, as the number of regimes to be identified by an RR algorithm increases, the ability to impart UBM knowledge via the manipulation of the evaluation database is reduced and the need to impart this knowledge via error cost weighting is critical.

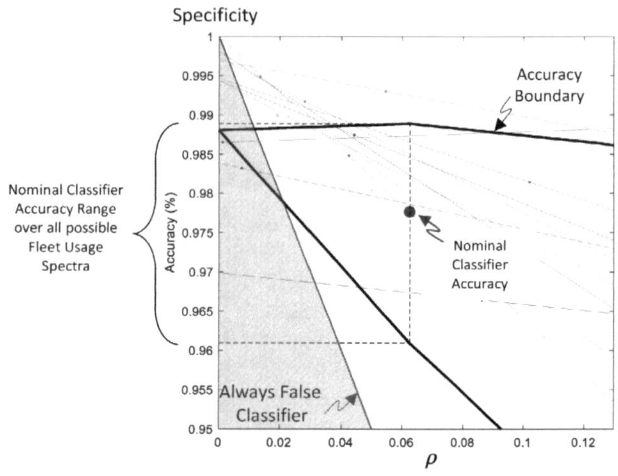

Figure 7: Deployed Classifier Accuracy Range for UH-60 IVHMU Onboard Regime Recognition Algorithm

It should also be noted that specificity will likely increase as the number of regimes increases. To explain this we refer back to the illustration in Figure 1. Specificity for any single regime is a measure of the correct classifications in the population in the right half of the circle. This population will grow significantly as the number of regimes is increased. This naturally increases the opportunity for TN classifications that can drive up specificity. Put another way, for any given unit of flight time there will be at least $N-2$ TN classifications assuming that the actual regime is missed. If the actual regime is correctly identified there will be $N-1$ TN classifications and one TP classification.

This phenomenon may lead to a desire to use metrics invariant to the correct classification of negative events. For this application, however, negative events are well defined maneuvers and we don't want to ignore FP errors. For example if severe regimes are missed it is important to know the extent to which the corresponding FP errors underestimate damage. We saw in the presentation of accuracy curves in the previous section that if the FP errors are classified as benign versus a different but similarly severe regime, such an error "costs" more and our analysis methods will highlight this difference via the error cost weighted accuracy score. FP errors are also important when regimes contribute to fatigue damage that would lead to the unnecessary replacement of costly components. In both such instances we would want to be able to impart this information via error costs to obtain an accuracy score that is more dependent on FP errors. Conversely, if the correct classification of negative events is not at all important for a particular application we still have a mechanism to impart this information via error costs to obtain a measure of informedness ($\rho_c = 0.5$), or to heavily favor recall, which is invariant to TN classification results (see Eq. (1)).

FP errors can also be captured in the precision metric given by,

$$P = \frac{TP}{TP + FP'} \quad (21)$$

which is a measure of correct classifications in the population contained in the inner circle of Figure 1 (i.e., the percentage of positive predictions that are correct). While this is an informative metric for the one-to-many analysis of individual regimes, when this metric is micro-averaged to obtain an overall measure of precision it collapses to the same quantity in Eq. (16).

We now use error cost weighting to impart general UBM knowledge that it is important not to miss autorotation maneuvers, pullouts, pushovers, and turns as these maneuvers are related to multiple failure modes for multiple safety-critical aircraft components. To impart this knowledge, we increase the cost of an FN error for these regimes. In Figure 8, we can see the effect is to move the operating points to the intersection with the trivial classifier. Any movement beyond this intersection would imply that recall is so important that a trivial Always True classifier should be employed.

It is also important not to overstate the occurrence of level flight as doing so may also lead to the underestimation of damage. Put another way, when a damaging regime is missed, we don't want the corresponding FP errors to accumulate in the level flight regime as this would lead to the greatest underestimation of aircraft damage. To impart this knowledge, we increase the cost of an FP error until the operating point is at the intersection with the Always False classifier. Again moving beyond this intersection would imply that not over-counting level flight is so important that we should simply say it never occurs.

For the remaining regimes we adjust the error cost weighting to place the operating points in the center of the operating range to obtain regime accuracy scores that are not biased by prevalence in the evaluation database (i.e., proportional to informedness).

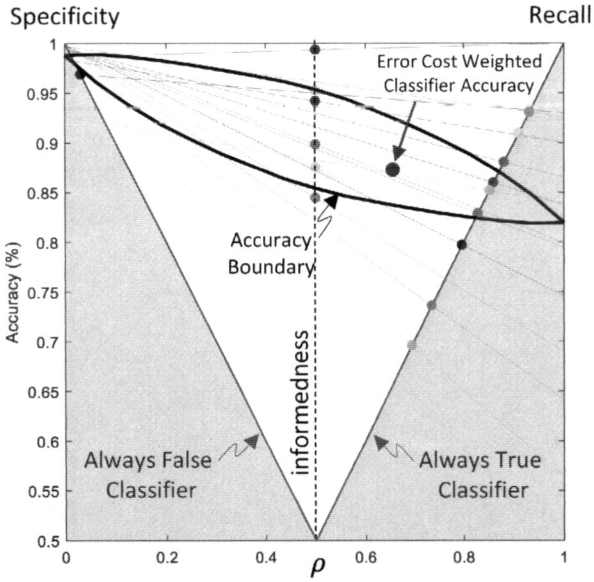

Figure 8: Regime Accuracy Curves and Error Cost Weighted Classifier Accuracy for UH-60 IVHMU Onboard Regime Recognition Algorithm

From Figure 8 we can see that imparting this general UBM knowledge moves the classifier operating point to the right resulting in a lower classifier accuracy score. Furthermore we know that since $\bar{\rho}_c$ is in the right half of this space, this new accuracy metric is more a measure of how likely damaging regimes will not be missed. As such a classifier with superior recall for damaging regimes would stand out. From Figure 6 we can see that without the error cost weighting a classifier that has superior recall for damaging regimes will not result in a significant change in accuracy since the makeup of the evaluation database keeps the operating point in a region where accuracy is highly dependent on specificity.

Confidence Bounds

With sufficient evaluation data, it is also possible to generate confidence bounds over the entire operating range. This is done by randomly selecting subsets of the data with the same maneuver prevalence as the whole dataset. For each subset, the individual regime accuracy curves are generated, the convex hull boundary is found, and the same error cost weighting is applied to obtain the classifier accuracy score. The results are shown in Figure 9.

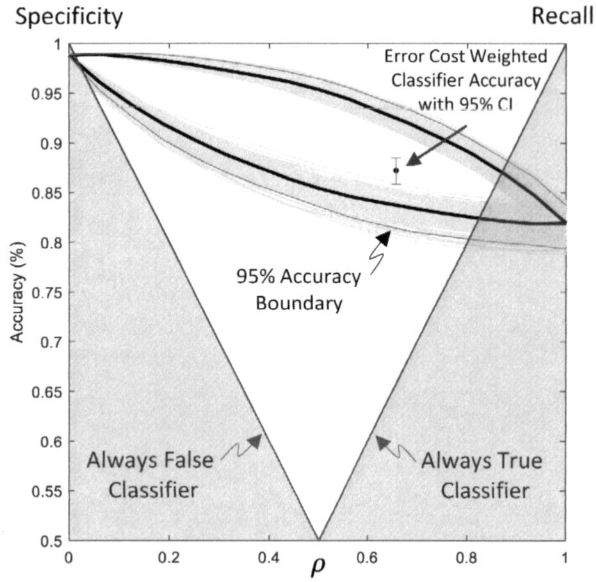

Figure 9: 95% Confidence Bounds for Classifier Accuracy Convex Hull and Error Cost Weighted Classifier Accuracy

Each convex hull accuracy boundary is plotted in light grey with the 95% confidence boundary given in red. The 95% confidence boundary for the error cost weighted classifier accuracy is shown with the red error bars.

Component Specific Knowledge

If a candidate RR algorithm is to be used as part of a UBM program, we would want to know how regime classification accuracy translates to accuracy in estimating component damage. In Ref. [1] is it specified that a candidate RR algorithm cannot under predict the fatigue damage fraction of life-limited components by more than 0.5%. This metric can be calculated directly from the RR algorithm output; however, in so doing this ignores the valuable information embedded in the categorical results. For example, it could be that this metric is satisfied, but only because of excessive FP errors. To ensure this information is not missed we use this analysis framework to impart component specific knowledge. By putting this metric into this context of accuracy curves we can provide a complete picture of important error characteristics.

To demonstrate this capability we selected the stationary swashplate. From Ref. [6] we know that approximately 41% of the fatigue damage to the stationary swashplate is caused by 45° left turns, followed by 39% for 45° right turns, and 20% from pullout maneuvers. This information is imparted via the weighting afforded through ω_i in Eq. (19) to convey the relative importance of each regime. The relative importance of Type I versus Type II errors is imparted via Eq. (8) to impart the knowledge that we only car about missing these regimes. We first show the nominal accuracy in Figure 10.

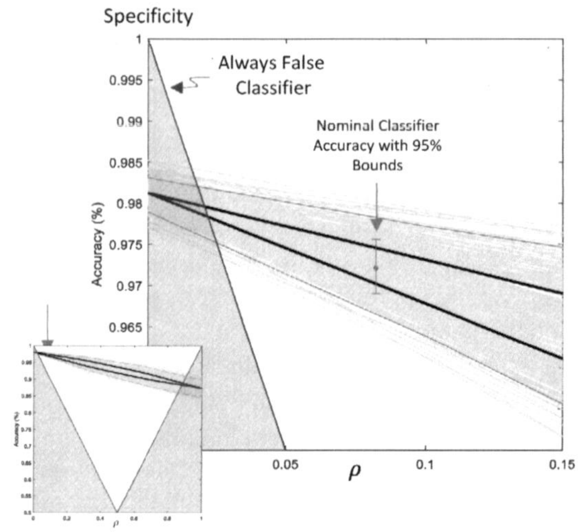

Figure 10: 95% Nominal Swashplate Accuracy

Given its location in this space we know that the nominal accuracy metric is a measure of how likely the candidate RR algorithm will not overestimate damage. The error cost weighted accuracy is shown in Figure 11.

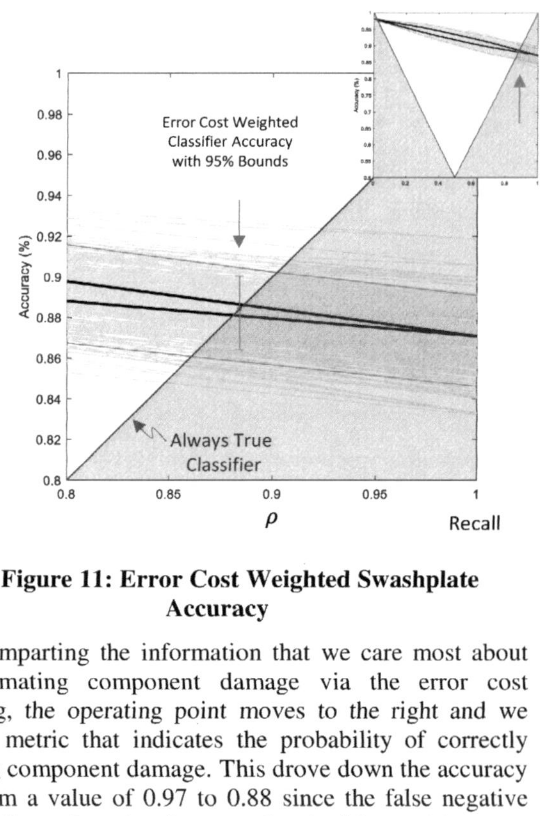

Figure 11: Error Cost Weighted Swashplate Accuracy

By imparting the information that we care most about underestimating component damage via the error cost weighting, the operating point moves to the right and we obtain a metric that indicates the probability of correctly capturing component damage. This drove down the accuracy score from a value of 0.97 to 0.88 since the false negative error rate for each regime is worse than its false positive error rate.

CONCLUSIONS

The analysis methods and associated visualization afforded through accuracy curves provide a context for important metrics including sensitivity, specificity, accuracy, and informedness. Presenting these metrics as a function of prevalence provides a complete picture of how a candidate RR algorithm will perform when it is deployed to the fleet to estimate a usage spectrum and provides an overall classifier accuracy metric that will highlight error characteristics important to a UBM program.

The ability to impart UBM specific information via error cost weighting, in either a general sense or specific to life limited components, will ensure that the reported classifier accuracy score is dependent on the types of errors that would lead to the underestimation of aircraft damage. As such, we know that if the primary objective of achieving an accuracy of 97% is satisfied, we also know that misclassifications err on the side of selecting more severe regimes. When comparing multiple candidate RR algorithms, this capability provides a single metric that highlights even the most subtle differences in error characteristics that are critical to the estimation of aircraft fatigue damage. The ability to include confidence intervals over the entire operating range will further facilitate the comparison of multiple RR algorithms to determine if there is a statistically significant difference between them.

Using accuracy curves we showed the dependence of the accuracy score on the prevalence of maneuvers in the FLS evaluation database and that for a multi-class classifier this dependence can result in a metric that is biased by classification errors not important to a UBM program. This was demonstrated on an RR classifier configured to reflect the existing UH-60 IVHMU onboard RR algorithms. The accuracy obtained from the FLS evaluation database was found to be sufficient per the ADS-79-HDBK handbook; however, the accuracy curves showed that this high score only reflected the fact that damaging regimes were not likely to be over-counted. Using error cost weighting the dependence on the makeup of the evaluation database was removed and replaced with general UBM knowledge to obtain an accuracy score that highlighted the ability of the RR algorithm to retrieve damaging regimes and not over-count benign regimes as both would lead to the underestimation of aircraft damage. This resulted in a significantly lower score since the classifier did not have strong recall capabilities for these critical regimes. Applying this same process to multiple candidate RR algorithms would allow developers to identify a solution that has the desired error characteristics.

We then imparted component specific information via error cost weighting to obtain an accuracy score that reflects the ability of an RR algorithm to estimate component damage. Such information can be imparted for multiple aircraft components and damage modes to obtain a collection of component accuracy scores, which can then be aggregated based on the relative importance of each component. In so doing the resulting metric is dependent on the classification errors most important to the estimation of aircraft damage. The associated visualization provide a convenient space to unpack this high level metric to understand what classification errors are driving the score.

Author contact: Jason Hull, jason.hull@spire-innovations.com, Jeffrey Monaco, jeffrey.monaco@spire-innovations.com, Mark Davis, mark.w1.davis@lmco.com, Raymond Beale, ray.j.beale.jr@lmco.com, Roberto Semidey, roberto.semidey@navy.mil, Mark Glucksman-Glaser, mark.glucksmanglaser@navy.mil.

ACKNOWLEDGMENTS

This work was conducted under a NAVAIR SBIR Phase I contract. We would like to thank the technical monitors, Mark Glucksman-Glaser, and Roberto Semidey for their support in this effort. We would also like to thank our corporate partners at Sikorsky Aircraft Corporation for their support in this work.

REFERENCES

[1] M. McCall, D. Stephan and J. Lackey , "Aeronautical Design Standard Handbook: Condition Based Maintenance System for U.S. Army Aircraft," U.S. Army Aviation Engineering Directorate (AED), 2013.

[2] C. Drummond and R. Holte, "Cost curves: An improved method for visualizing classifier performance," *Machine Learning,* vol. 65, pp. 95-130, 2006.

[3] D. Powers, "Evaluation: From Precision, Recall, and F-Measure to ROC, Informedness, Markedness and Correleation," *Journal of Machine Learning Technologies,* pp. 37-63, 2011.

[4] M. Sokolova and G. Lapalme, "A systematic analysis of performance measures for classification tasks," *Information Processing and Management,* vol. 45, pp. 427-437, 2009.

[5] A. M. Dere, "Flight regime recognition analysis for the Army UH-60A IMDS usage," Naval Postgraduate School, Monterey, CA, 2006.

[6] R. J. Beale, M. W. Davis and P. Swindell, "Achieving Usage Based Maintenance with HUMS Regime Recognition," in *AHS 71st International Forum*, Virginia Beach, 2015.

In-Flight Tuning: Flight Test Results and 2nd Generation System

Dr. Richard Hunt
Systems Engineer
Helitune Ltd.
Torrington, UK

Dr. Dominic Southgate
Project Manager
Helitune Ltd.
Torrington, UK

Douglas Graham
General Manager
Helitune Ltd.
Torrington, UK

Dr. Daniel Fuerst
Program Manager
ZF Luftfahrttechnik GmbH
Kassel-Calden, Germany

Dr. Uwe T. P. Arnold
Director Development Rotor Components
ZF Luftfahrttechnik GmbH
Kassel-Calden, Germany

ABSTRACT

Helicopter rotor systems feature adjustable elements, such as pitch links, balance weights and/or trim tabs that allow the vibration levels in the aircraft to be reduced. The required adjustments are normally determined using Rotor Track and Balance (RT&B) equipment on dedicated flights. However, these adjustments are a compromise intended to provide a solution for only the most common flight regimes and aircraft loading conditions. This work presents the development of an In-Flight Tuning (IFT) System, featuring Smart Pitch Rods (SPR™s) that enable continuous balance adjustments and optimization for any aircraft condition. Initial testing of a Flight Trial System on a CH-53G test bed showed that a noticeable effect on aircraft vibration could be achieved using the SPR™s in open-loop control and continuous vibration minimization was possible using a closed-loop control algorithm. Additionally, the flight trial showed that the IFT System could give improvements over the traditional RT&B compromise and also showed reductions in vibration to within aircraft limits after a blind blade exchange. A production IFT System is now in development, featuring redesigned components and an extended set of functions that are described in the final section.

NOTATION

1/rev	Once per revolution
1R	1st harmonic of main rotor vibration
1T	1st harmonic of tail rotor vibration
ACU	Actuator Control Unit
DAL	Design Assurance Level
DAT	Digital Audio Tape
HHU	Hand Held Unit
IBC	Individual Blade Control
IFT	In-Flight Tuning
MPU	Magnetic Pick-Up
MR	Main Rotor
NFAC	National Full-Scale Aerodynamic Complex
PCB	Printed Circuit Board
RT&B	Rotor Track & Balance
RTS	Return to Service
SPR™	Smart Pitch Rod
TR	Tail Rotor
VMU	Vibration Monitoring Unit
ZFL	ZF Luftfahrttechnik GmbH

Presented at the AHS Airworthiness and HUMS Technical Meeting, Huntsville, Alabama, USA, February 21-22, 2018.

INTRODUCTION

Helicopter rotor systems commonly feature a number of adjustable elements that can be tuned in order to minimize the vibration levels that are observed within the aircraft. The parameters available for adjustment are normally the length of the pitch link, the addition or subtraction of balance weights and/or bending of trim tabs that are attached to the trailing edge of the blades. This set of possible adjustments is used to compensate for the 1R vibration and blade track misalignments that result from the full spectrum of aerodynamic and mechanical dissimilarities between aircraft. These may be derived from geometric, elastic or mass deviations between the blades as well as variations in the airframe geometry and stiffness. The appropriate adjustments to each parameter are generally determined using Rotor Track and Balance (RT&B) equipment that measures the vibration levels at a number of locations on the airframe and the relative track of the blades at a number of pre-defined flight conditions (airspeeds).

The "optimum" set of these adjustable elements is always a compromise, because they vary across the flight conditions and with aspects such as the configuration and loading of the aircraft. The RT&B process requires multiple flying hours and may still result in high vibration levels when flying in conditions outside of those used to generate the solution. In light of this, ZF Luftfahrttechnik GmbH was contracted in 2010, with the support of Helitune, to develop and flight-test a system that would allow the continual adjustment of the pitch links to enable In-Flight Tuning (IFT) to minimize aircraft 1R vibration. The resulting prototype featured six Smart Pitch Rods (SPR™) installed on a CH-53G test bed.

This paper briefly discusses the Flight Trial system and the results obtained during the flight test campaign before discussing the current efforts to develop a production IFT System.

FLIGHT TRIAL SYSTEM

The IFT Flight Trial System was developed following successful full-scale Individual Blade Control (IBC) system wind tunnel tests in the National Full-Scale Aerodynamic Complex (NFAC) at Ames. These tests were the culmination of a long-term collaborative effort between NASA, U.S. Army, Sikorsky Aircraft Corporation, and ZF Luftfahrttechnik GmbH (ZFL) to demonstrate the benefits of IBC for a UH-60A rotor (Ref. 1). During this test campaign ZFL had the possibility to demonstrate the feasibility of removing 1R imbalances from the rotor system using active pitch links (Ref. 1, Ref. 2) and hence to evaluate the concept of In-Flight Tuning.

The IFT Flight Trial system that followed used six electrically actuated Smart Pitch Rods (SPR™s) to replace the fixed pitch links on the CH-53G. The length adjustment of each SPR™ was commanded individually by the IFT Commander which also monitored and recorded the health and status of the SPR™s. Depending on the mode of operation of the IFT Flight Trial system the IFT Commander processed length adjustment commands received from different interfaces.

The IFT Flight Trial System was installed on a CH-53G testbed of the German Armed Forces in 2012 and the flight test campaign was performed at the German Military Flight Test Center in Manching. The tests were designed to prove the IFT concept in a realistic environment and have successfully demonstrated autonomous IFT operation based upon closed loop control algorithms.

FLIGHT TRIAL SYSTEM ARCHITECTURE

The architecture of the IFT Flight Trial System is shown in Figure 1 below. The system elements in white are retained in the production system although most have been redesigned as described later in this paper. Details of the Flight Trial System installation can be found in Ref. 3 and Ref. 4.

The Flight Trial System had multiple modes of operation allowing the SPR™ adjustments to be in closed-loop control (using a balancing algorithm running on the dSPACE sub-system), under direct open-loop control (from dSPACE or the IFT Hand Held Unit (IFT HHU)) or controlled from a Helitune carry on RT&B system (RotorTuner 2000). The operator could select the mode of operation from the IFT Hand Held Unit interface depending on the test being performed. Initially the SPR™s were under manual control with their positions being manually commanded and their movement monitored on the IFT Hand Held Unit and the dSPACE sub-system. Both, the dSPACE sub-system and the Rotortuner 2000 were used to evaluate the impact of the SPR™ adjustments on the vibrations. Primary focus was put on the addressed 1R vibrations but the potential impact on higher harmonics was investigated, too.

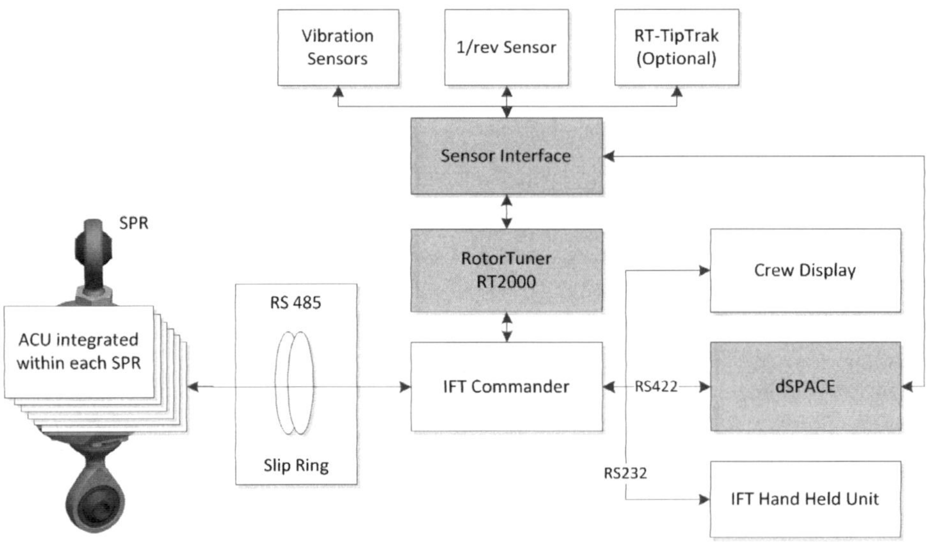

Figure 1: IFT Flight Trial System Architecture

The Rotortuner 2000 allowed the principle of in-flight RT&B to be investigated in a semi-active closed loop configuration and finally the dSPACE sub-system allowed tests to be carried out on the closed loop control algorithm under fully autonomous operation. In all modes of operation, the Crew Display provided a system "Unarm" control which allowed the pilot to disable any SPR™ adjustments. Flight Trial operators station is shown in Figure 2.

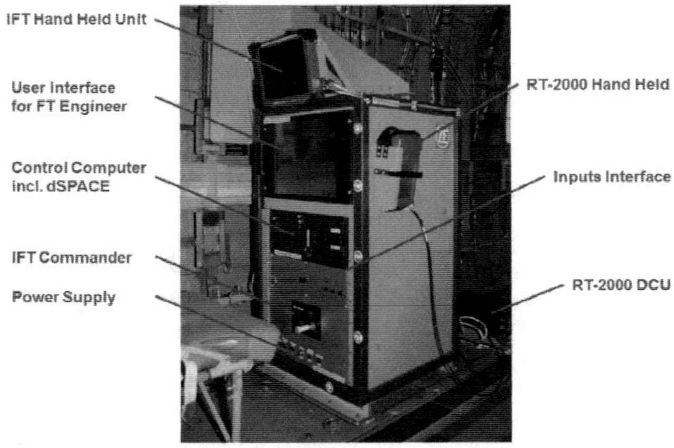

Figure 2: IFT Flight Trial System Operators Station

CLOSED LOOP CONTROL MODE

Closed-loop control was provided by a commercial-off-the-shelf dSPACE system which provided a simple platform for the development of the closed-loop algorithm. It took inputs from the sensors and SPR™ positions from the IFT Commander to calculate SPR™ length adjustments resulting in pitch adjustments to minimize the 1R vibrations. These two sub-systems were the primary active components during closed-loop control of the IFT system. Unlike the RT&B process, the SPR™ adjustment calculated by dSPACE system in closed-loop operation were specific to the current flight condition and 1R vibration level, so provided a better balance than the typical compromise setting that can be achieved during today's traditional RT&B. The IFT Hand Held Unit was used to switch operating modes but was otherwise passive while under closed-loop control.

The SPR™s featured mechanically limited authority during flight operation which for the Flight Trial campaign was set to 10 clicks (approx. 1.32 mm length adjustment). Further adjustment of the SPR™ could be made manually on ground when needed for RT&B or aircraft rigging, following removal of so called hard-stop pins, to give the SPR™s a similar total adjustment range to the original pitch links. This manual adjustment mechanism has been replaced with full electrical adjustment in the production IFT System.

RT&B MODE

The Helitune RT-2000 RT&B system was included within the architecture of the Flight Trial system (cf. Figure 1) to allow calculation of a traditional RT&B compromise adjustment while the system was in use. This allowed a verification of the measurements provided through the closed-loop control hardware with those calculated using the RT-2000 which had been in use on the CH-53 within the German Armed Forces for many years. The RT-2000 software was modified to allow the calculated adjustments to be applied directly to the SPR™s in-flight via a connection to the IFT Commander, which provided confirmation of the effectiveness of the compromise RT&B adjustment without requiring ground based adjustments and this important functionality has been retained as the "Return to Service" mode of operation in the production system.

The IFT Commander and the RT-2000 were the primary active components during RT&B mode of the system. Again, the IFT Hand Held Unit displayed the SPR™ positions and status, and allow the operating mode to be switched.

DIRECT CONTROL MODE

In this mode of operation, the IFT Hand Held Unit or the dSPACE sub-system was used to set the SPR™ adjustments in open-loop mode and to introduce "disturbances" for testing (e.g. disabling the use of an SPR™ to simulate an SPR™ failure or adding an imbalance to simulate sudden blade damage). Moreover, this mode was used to evaluate and confirm the assumed impact of single or multiple SPR™ adjustment on the 1R imbalance. Direct control was useful for flight testing but is not available in the production system.

FLIGHT TEST RESULTS

Figure 3 shows the testbed aircraft during one of the early test flights.

Figure 3: CH53G test aircraft used for IFT Flight Trials
[Photograph Courtesy of Alex Beck]

The following sections briefly summarize major results presented in Ref. 3.

Open Loop tests

The first phase of the test flights involved commanding the SPR™s in open-loop mode. SPR™ adjustments were made to the rotor to assess their impact; primarily on the 1R lateral (measured on the upper housing of the main gearbox) and vertical (measured in the nose avionics bay) vibrations. All other n/rev harmonics and the blade tip path were continuously monitored to ensure there were no unwanted side-effects.

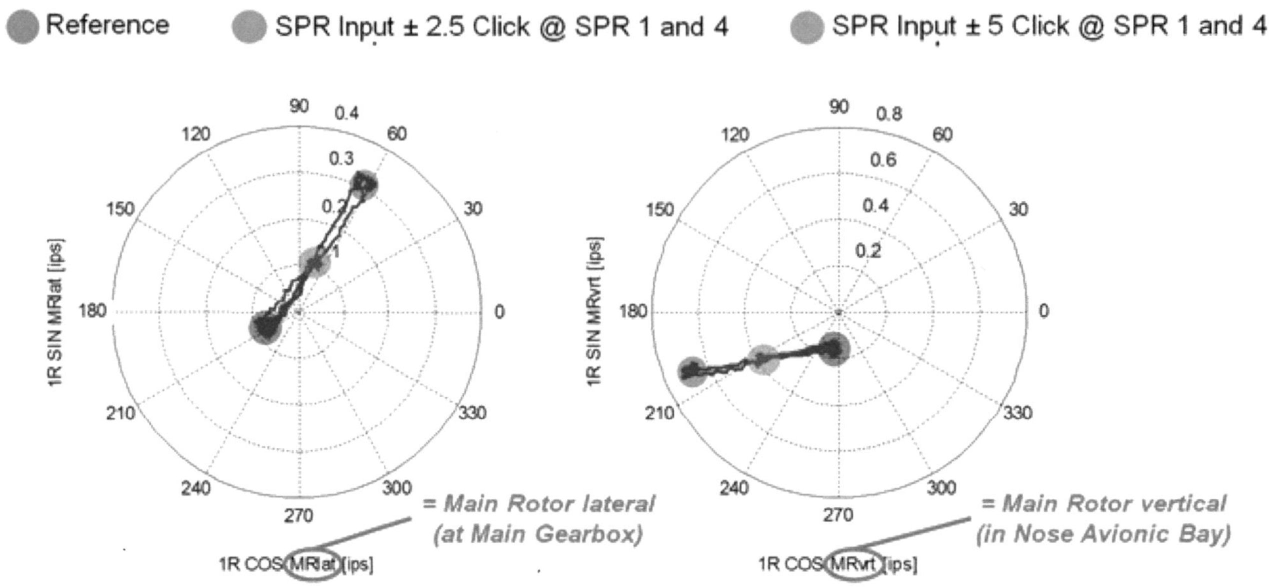

Figure 4: Effect on MR Lateral and Vertical 1R of ±2.5 and ±5 click SPR™ displacements simultaneously applied to opposite blades

Figure 4 shows the response of the 1R main rotor lateral and vertical balance to SPR™ adjustments of two opposite SPR™s, in the polar domain. This test was carried out at 130 kts straight and level flight. The reference vibration with the SPR™s at their default mid positions, is shown as a blue circle. First an adjustment of +2.5 clicks and -2.5 clicks was made to SPR™s on opposite sides of the rotor head. This was then increased to ±5 clicks and then reduced back to default positions. The blue trace shows how the 1R main rotor balance changed during the test reaching the green circle at 2.5 clicks and the red circle at 5 clicks. The balance change was basically linear with SPR™ adjustment and repeatable. In the polar plots, the concentric rings represent the vibration amplitude in ips (up to 0.4 ips for lateral and 0.8 ips for vertical)

Closed Loop tests

Following successful completion of the open-loop testing in which the system showed a linear response to SPR™ adjustments with no anomalous behavior, the Flight Trial moved on to testing the closed-loop performance which was the main area of interest. The initial tests were designed to examine the effect on the reference level of vibration when the IFT closed-loop algorithm was enabled. During straight and level flight, the reference level of vibration was measured with the IFT System under direct control from the Hand Held Unit and SPR™s at their default positions. The IFT System was then switched to dSPACE closed-loop control and then back to manual control with the SPR™s at their default positions. At this early stage, just 2 adjacent SPR™s were enabled which limited the authority of the system but also showed the system could operate effectively with the reduced authority.

In Figure 5 the IFT closed-loop control is operational for approximately 3 minutes at 110knots straight and level flight (green trace). The reference periods before and after enabling closed-loop control are shown in red and the transition period is shown in blue. The figure clearly shows the reduction in 1R vibration for the lateral and vertical while the system is operating in closed loop. It can be seen that even with just 2 adjacent SPR™s active, the required adjustment is less than 3 clicks.

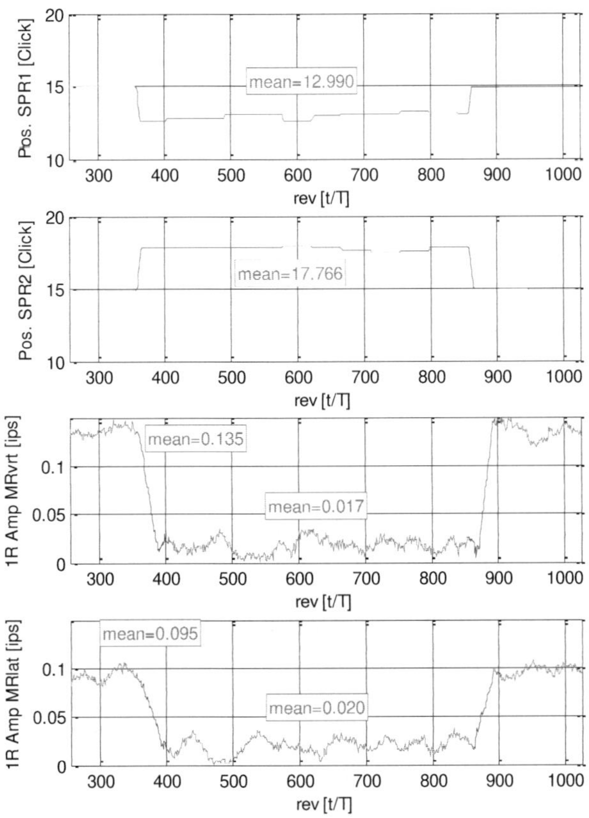

Figure 5: Closed loop control sequence - top: SPR™ positions, bottom: 1R vibrations

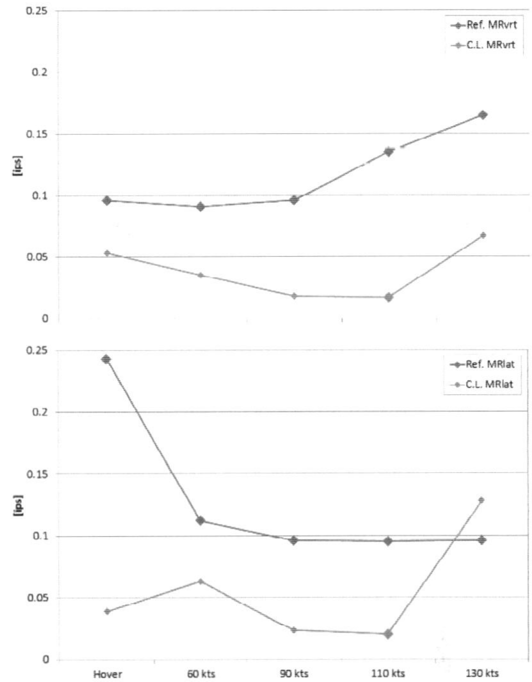

Figure 6: 1R vibration levels recorded with (green) and without (red) closed-loop control at various flight conditions

IFT vs RT&B Compromise

The testing then moved on to examine the IFT closed-loop performance against the traditional RT&B compromise adjustments. One of the key objectives of the IFT System was to provide a smoother flight across all flight conditions than could be achieved with traditional RT&B. In this test the aircraft completed the traditional RT&B so that the vibration was within limits across a range of flight conditions. Then at each flight condition, the vibration levels were measured with and without the IFT System active.

The results presented in Figure 6 show that, with the exception of lateral at 130knots, the IFT close-loop control (green) provided significantly lower 1R vibration than the reference level (red). At 130 knots, the IFT System has sacrificed a small increase in lateral vibration to provide a large reduction in the vertical vibration, resulting in a lower overall vibration level. Tests were also performed with different loading and CG conditions and at different altitudes and in all cases the IFT System gave significant reductions in overall 1R vibration.

Blade Exchange Test

One of the final tests to be performed involved a blind blade change. In military operational environments, aircraft availability can be severely hampered by the time required for RT&B when a blade needs to be exchanged so the ability to compensate for the vibration caused by a blade exchange would be beneficial. In this test the vibration levels were first measured at the standard RT&B flight conditions without the IFT System to give a reference. Then a rotor blade was removed and replaced with a spare blade delivered from a central depot. The length of the corresponding SPR™ was manually adjusted according to the pre-track information that came with the spare blade. No other changes were made to the SPR™s or the aircraft.

Following replacement of the blade, the aircraft was flown at the same RT&B flight conditions with and without the IFT System enabled.

Figure 7 shows the 1R amplitude for the main rotor lateral and vertical vibration. Before the blade exchange the aircraft was within the RT&B limits for the CH53G with the IFT System turned off (purple). After the blade exchange (Red), the vibration levels had increased to well above the RT&B limits of 0.3ips and 0.2ips for lateral and vertical respectively. However, when the IFT System is turned on (green), it is able to reduce the vibration levels once more below the RT&B limits and in most cases below the pre-blade change vibration levels.

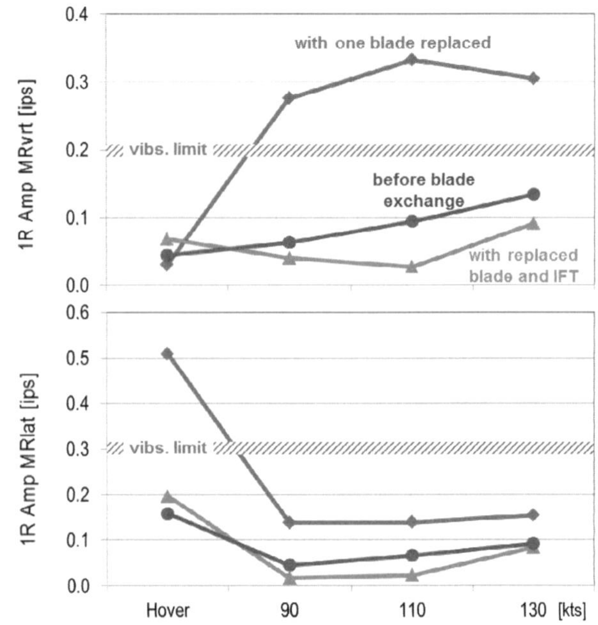

Figure 7: Compensation of 1R vibrations caused by a replaced blade

PRODUCTION SYSTEM

SYSTEM ARCHITECTURE AND DEVELOPMENTS

Under a governmental contract awarded by the German Armed Forces to ZF Luftfahrttechnik GmbH, a second generation IFT System is now under development with the support of Helitune Ltd., building on the successful flight trial program, to create and certify a production-ready version of the IFT System. The main components of this system are as follows (see Figure 8):

Overall System

The control system hardware and especially their integration used for the Flight Trial system were not fully suitable for an on-board system integration due to its bulk (see Figure 2) and so a redesign of the architecture and corresponding system elements was necessary. The IFT Junction Box, IFT Commander and IFT VMU perform the functions retained for the production system and previously performed by the components in the Flight Trial operators station shown in Figure 2 but will be distributed in the aircraft fuselage based on available and appropriate installation space.

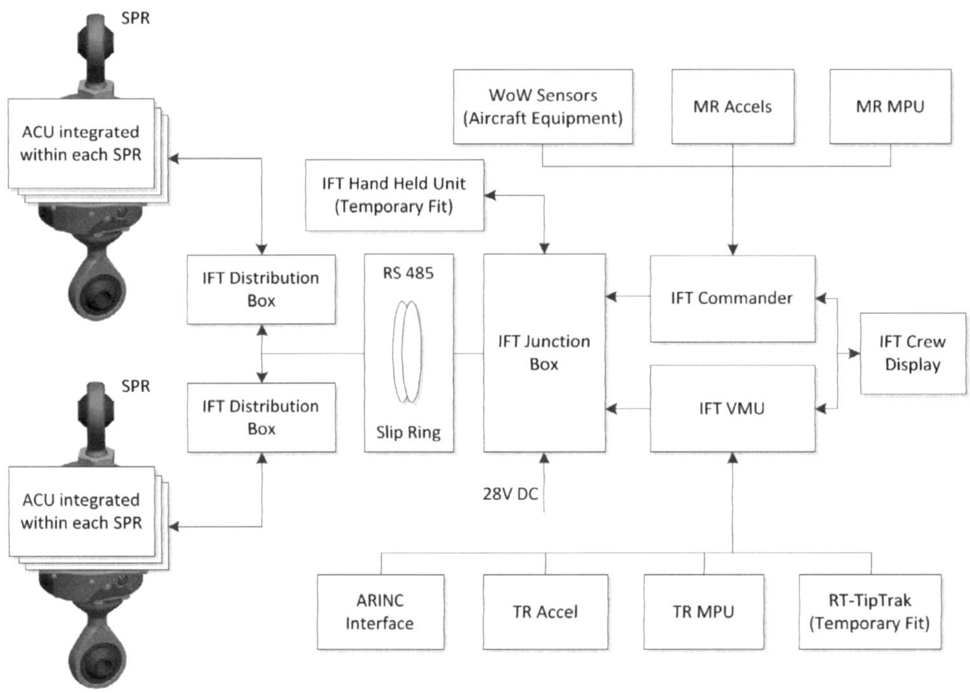

Figure 8: IFT System Architecture

IFT Commander

The IFT Commander hardware architecture is a single micro-controller based PCB which has been specifically designed by Helitune to support the IFT System requirements (Figure 9). It interfaces to the main rotor sensors, Weight-on-Wheels sensors, SPR™s, IFT Crew Display and IFT VMU. The sensor interface allows the IFT Commander to measure the 1R vibration of the main rotor from the accelerometers and 1/rev sensor (MPU). Dual lateral and vertical accelerometers have been included on the main rotor to provide a mechanism for identifying sensor or mounting failures.

Figure 9: The IFT Commander for the production IFT system

The IFT Commander takes input data from the SPR™s, main rotor accelerometers, 1/rev sensor, and the weight-on-wheels sensors. The connection to the SPR™s (via the IFT Junction Box) allows the IFT Commander to gather position, status and health information and to command the SPR™s to new positions.

In Normal Flight Mode operation the IFT Commander autonomously applies adjustments calculated using the closed-loop control algorithm that was previously developed by ZF Luftfahrttechnik GmbH and implemented on the dSPACE sub-system for the Flight Trial system. Closed loop control starts when the aircraft is airborne (indicated by the weight-on-wheels sensors) to avoid any unwanted interaction with Ground vibrations, and stops when the aircraft lands. The data storage functionality of the flight trial based IFT Commander has been moved to the IFT VMU. This has been driven by the need to execute the core closed loop algorithm within a higher-integrity (software) component (the new IFT Commander) to satisfy the System Safety Assessment results.

In Return to Service Mode the IFT Commander provides an interface between the IFT VMU and the SPR™s in order to provide SPR™ position and health information and to apply RT&B adjustments.

The IFT Commander has a discrete I/O interface to the IFT Crew Display which allows the IFT Commander to indicate the system status to the aircrew and for them to command the system between the ARMED and UNARMED states. In the UNARMED state, the closed-loop control is inactive and all SPR™s are commander to their default mid-position.

All system measurements and status information from the IFT Commander is shared with the IFT VMU over a broadcast interface allowing it to be used in the Return to Service mode, analyzed for anomalies and stored for further off-line analysis.

IFT VMU

The IFT VMU is based on the existing Helitune HT VMU that provides RT&B functions and data recording. It interfaces with the IFT Commander, IFT Crew Display, tail rotor sensors, aircraft ARINC data bus and blade tracking camera via the IFT Junction Box.

Data from the tail rotor accelerometers and 1/rev sensor is combined with data from the main rotor accelerometers, 1/rev sensor and blade tracking camera (RT-TipTrak or Linescan camera) in order to provide a complete on-board RT&B solution with the advantages afforded by in-flight adjustment of the SPR™s on the main rotor.

The IFT VMU also records main and tail rotor 1R/1T measurements and time domain vibration data periodically during Normal Flight mode or when initiated from the IFT Crew Display. On the ground, it outputs maintenance indications to the IFT Crew Display located in the cockpit and allows the maintenance actions to be acknowledged.

The IFT System currently uses 6 of the 16 accelerometer input channels and 2 of the 8 tachometer input channels available on the IFT VMU which significant room for extending the capabilities of the system into monitoring of drive train and gearbox vibration.

A development over the Flight Trial system is the recording of aircraft data from an ARINC interface if available. This will allow aircraft usage metrics to be calculated and may in future be used for flight regime detection.

IFT Crew Display

The IFT Crew Display has been significantly developed from the Flight Trial system to provide a textual display with illuminated indicators, to provide needed compliance with Night Vision Imaging Systems and enable pilot interaction with the IFT system (Figure 10). The IFT Crew Display communicates with the IFT Commander and IFT VMU giving the pilot the capability to arm/unarm the system, set event markers in case of unusual aircraft behavior, view SPR™ status information and review maintenance codes transmitted by the IFT VMU.

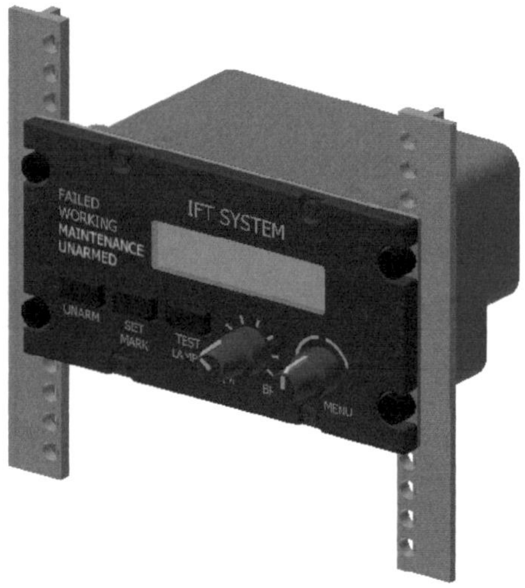

Figure 10: The IFT Crew Display for the production IFT system

IFT Junction Box

The IFT Junction Box was added to the production system to provide data communications between each of the on-board IFT sub-system components and power distribution with circuit protection. The IFT Junction Box also provides connections for temporary fit equipment including the Helitune RT-Backup auxiliary power supply for ground operation purposes, RT-TipTrak or Linescan blade tracking cameras and the IFT Hand Held Unit.

The IFT Junction Box provides separate connections to the IFT Hand Held Unit for the Return to Service and Ground Support modes and ensures the IFT Commander is automatically disconnected from the SPR™s in Ground Support mode (cf. Section "System Operation"). The IFT Junction Box design is specific to the CH-53 integration, owing to the power distribution requirements of the specific aircraft configuration. Integration onto different aircraft platforms may result in different configurations of the IFT Junction Box.

IFT Hand Held Unit

The IFT Hand Held Unit of the production system is based on the existing Helitune Flitepad and provides the main user interface when operating in Return to Service mode, Ground Support mode or SPR Maintenance mode (i.e. not during normal operation). It is a ruggedized tablet device with an 8" touch-screen display and magnesium alloy enclosure (Figure 11).

The IFT Hand Held Unit is connected to the system using specific cables/connections depending on the mode of operation. In SPR maintenance mode the IFT Hand Held Unit is connected directly to a single SPR™ for configuration, test and calibration activities. Whereas in Return to Service or Ground Support mode the IFT Hand Held Unit is connected to the relevant input on the IFT Junction Box.

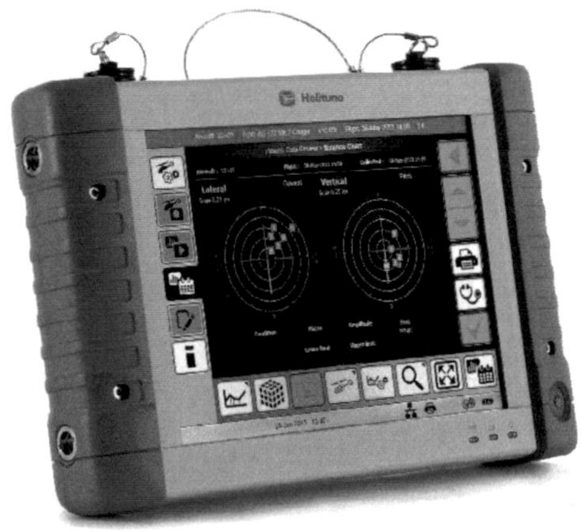

Figure 11: The IFT Hand Held Unit tablet device for the production IFT system

SPR™s & Slip ring

As for the flight trial system the SPR™ replaces the original fixed-length pitch link of the helicopter. It acts as an electro-mechanical actuator, which provides the capability to adjust the length of the pitch link with an authority, adjustment speed, and load capability suitable for the application to the CH-53G/GS/GE platform. The production system was derived from the flight test version by introduction of a few modifications. The authority of the electrical SPR™ adjustments available during Normal Flight Mode and Return to Service Mode operation, called dynamic SPR™ authority, was tailored such that an appropriate operation based on the achievable balance change is assured, but the impact on the aircraft safety remains at an acceptable level for functional failures related to hard- and software portions of the equipment. The investigations yielded to a design with a dynamic SPR™ authority of 5 Clicks.

As motivated before, the flight test version provided an additional manual length adjustment within a range similar to the current pitch links. This could be used for larger adjustments during aircraft rigging, application of blade pre-track adjustments or if larger adjustments would become necessary during Return to Service Mode operations. To simplify such larger adjustments and to reduce the likelihood of human error the production system SPR™ is modified to provide a fully electrical so-called Pre-Set adjustment capability available in Ground Support or SPR Maintenance Mode operation. The Pre-Set adjustment range is chosen such that the necessity for a mechanical length adjustment (which is still possible) of the SPR™s during aircraft rigging or Return to Service Mode operation is improbable. The following Figure 12 shows the production version SPR™.

The adjustment authority available during Normal Flight Mode operation is still limited safely by mechanical means, now called Pre-Set Lock Pin. If the Pre-Set Lock Pin is installed the SPR™ has an electrical adjustment authority of 5 Clicks. When the Lock Pin is unscrewed the complete Pre-Set adjustment range is available during Ground Support Mode or SPR Maintenance Mode operation on ground. For independent verification of the electrically applied Pre-Set adjustments (commanded and monitored by the IFT Hand Held Unit) the SPR™ features two mechanical scales. These can be used by the operator to verify that the commanded and electrically realized adjustments were accomplished correctly. Additionally to the above described design changes the SPR™ surface treatments were changed to REACH compliant procedures.

Figure 12: SPR™ for the production IFT system (from left to right: Pre-Set Lock Pin installed, Pre-Set Lock Pin dismounted, mechanical scale to verify Pre-Set adjustments)

The additional electrical slip ring (cf. Ref. 4) for power and signal transfer to the SPR™s is also retained. The existing design is revised to provide full compliance to all environmental qualification requirements of the production system and the solution still avoids interference with the existing electrical installation which provides power for the blade tip lights and folding circuity.

SYSTEM OPERATION

Normal Flight Mode

When power is applied to the IFT System, it automatically enters the ARMED Normal Flight mode in which the IFT Commander uses a closed loop algorithm to control the SPR™ adjustments. Initially on the ground, the IFT Commander checks the health status of the SPR™s and the connection status of the IFT HHU. If the IFT HHU is not connected, it proceeds to Normal Flight operation, otherwise it switches to Return to Service operation.

Closed loop control only starts when the aircraft is airborne and ARMED. If the flight crew feel there is a problem with the system, it can be stopped at any time by via an ARM/UNARM button on the IFT Crew Display. When UNARMED, the system commands all SPR™s to their default mid-positions which will provide the compromise vibration levels from the last RTS flight. Closed loop control also stops if a persistent failure of the system is detected by the IFT Commander continuous health checks (e.g. 4 or more malfunctioning SPR™s) or a landing is detected.

The IFT Commander uses two lateral and two vertical accelerometers to continuously measure the main rotor lateral and vertical 1R balance data. If the independent balance measurements fall within a pre-defined vector difference limit, the measurements can be used to calculate

SPR™ adjustments that will reduce the 1R vibration. Otherwise the system records an error. If the error persists, the system will enter a default state in which the SPR™s return to their default positions and a maintenance action will be recorded by the IFT VMU.

In order to be robust to SPR™ failures, the IFT Commander is able to automatically adjust the SPR™ commands to compensate for SPR™s that are malfunctioning. The system will continue operating providing at least 3 SPR™s are healthy by redistributing the SPR™ adjustments from unhealthy SPR™s to healthy ones. Although the total authority of the system will be reduced, the Flight Trial showed that this redistribution was able to give the same reduced level of 1R vibration. When the aircraft lands, the maintenance light on the Crew Display is illuminated if any SPR™s are faulty and the Crew Display can be used to identify the SPR™ and the fault code
Figure 13 shows how the autonomous Normal flight mode of operation is a continuous loop.

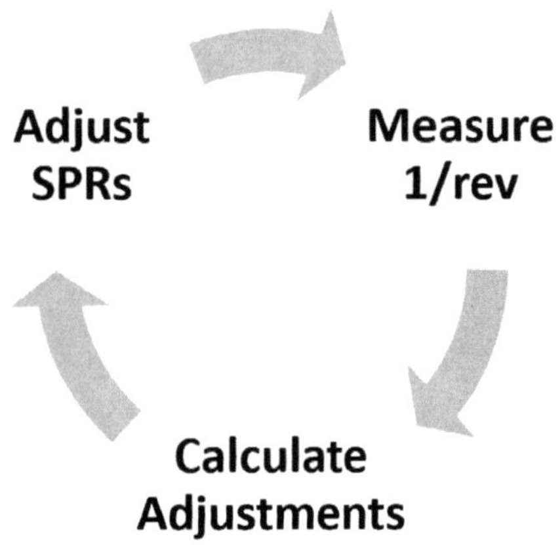

Figure 13: Closed-Loop Control Cycle

During Normal operation, the IFT Commander regularly sends the data it has gathered/calculated to the IFT VMU where it is stored on an internal SSD. This data includes the calculated 1R balance, the weight-on-wheels status, the SPR™ status and position, and the commanded SPR™ positions. In addition to storing data transmitted from the IFT Commander, the IFT VMU continuously records 1.6kHz sampled time domain vibration data (or up to 51.2kHz in 10second bursts) from the main and tail rotor accelerometers, data transmitted over ARINC-429 from the air data computer, and allows the aircrew to set event markers in the datastream using a button on the IFT Crew Display interface. This last function allows the aircrew to mark a point in time so that the data can be analyzed offline. The ability to store time domain data effectively enables the system to be used as a DAT recorder for offline analysis and development of condition indicators.

The IFT System has two performance monitoring functions in addition to the continuous health monitoring functions. During normal operation, the IFT VMU monitors the measured vibrations and if they are persistently high, it recognizes that the system performance has degraded (e.g. due to creeping blade changes that have unilaterally eaten up the adjustment margin). A second monitoring function compares the current flight SPR™ adjustments with a reference flight. If the current flight adjustment profile is too different to the reference flight this may indicate a change to the rotor system due to a mechanical fault. Both functions raise a maintenance indication which is reported via the IFT Crew Display when the aircraft is on ground.

Return to Service Mode

Return to Service mode replaces the traditional RT&B process on the CH-53 which is currently carried out with Helitune Carry-on equipment. RT&B for the main rotor is enabled using the SPR™s, allowing RT&B solutions to be applied mid-flight which can reduce the total number of flights required to balance the aircraft rotors. The system also has RT&B capability for tail rotor balancing, however, adjustments must be made manually.
In this mode, the IFT Commander remains the interface to the SPR™s and still broadcasts the SPR™ position, status and health to the IFT VMU. However instead of using the IFT Commander closed-loop algorithm to calculate SPR™ adjustments, they are calculated on the IFT Hand Held Unit and applied to the IFT Commander via the IFT VMU. The IFT Commander switches to Return to Service mode as soon as the IFT Hand Held Unit is connected to the IFT Junction Box.
If the Ground vibration is over limits, a solution is calculated manually by the operator using the IFT Hand Held Unit and providing the SPR™s have sufficient dynamic SPR™ authority, the solution can be applied and validated without the need for a second ground run. The target SPR™ positions are then transferred to the SPR™ "Pre-set position" using the Ground Support mode, so that the SPR™ has the maximum available dynamic authority for the remaining RTS flights.
Having balanced the aircraft on the ground, data is collected at the predefined Flight conditions with the SPR™s in their default mid-positions. If a solution is available that can be applied within the limited dynamic SPR™ authority, then this can be applied in-flight and the effect validated without the need for a dedicated check flight. If a solution is not available within the dynamic authority of the SPR™s, a compound adjustment using the larger SPR™ Pre-set authority and Tabs can be calculated for application on the ground. In this case a validation flight is required.
In either case any residual adjustment of the dynamic SPR™ position is transferred to the pre-set SPR™ position adjustment to enable the SPR™s to be operated with the maximum dynamic authority available for the Normal Flight mode operation.

Ground Support mode

Ground Support Mode is used to command the Pre-Set SPR™ positions that have a larger authority than the limited dynamic authority. The IFT Hand Held Unit is connected to a dedicated connector on the IFT Junction Box which causes the IFT Commander to be disabled (to prevent a conflict on the SPR™ communications) and directly connects it to the SPR™s. This allows the SPR™s to be at the optimum mid-point of their dynamic authority ready for normal operation. As explained the Pre-set SPR™ adjustment is protected by a mechanical Pre-Set Lock Pin which must be removed before the Pre-set position is changed. The Ground Support Mode application guides the user through each step of the SPR™ configuration using a wizard based system with instructions and confirmation checkpoints.

SPR Maintenance mode

SPR maintenance mode is carried out with a single SPR™ connected to the IFT Hand Held Unit and is intended to be a workshop activity but could be carried out on aircraft if necessary. It provides a full set of commands that allow the operator to test the SPR™s, calibrate them, set the address and update the SPR™ software.

SAFETY AND CERTIFICATION CONSIDERATIONS

SPR™ Authority

The SPR™ authority is deliberately small to ensure that even the worst case failure would not lead to unsafe operation of the aircraft. In the Flight Trial system four authorities were available ranging from 5 clicks to 10 clicks. It was found that, for normal operation an authority of 5 clicks is sufficient to optimize the 1R balance. A mechanical Pre-Set Pin within the SPR™ prevents it moving outside the defined authority during Normal operation but can be removed to provide a larger authority for applying ground adjustments resulting from the RTS process.

Design Assurance Level

The Design Assurance Level for the software on the IFT Commander is derived from the effect of the maximum vibration imbalance that could be caused by a set of completely dysfunctional and worst case misplaced SPR™s. The IFT Commander has direct control of the SPR™s and is responsible for suspending the system operation when the system is Unarmed by the pilot and for these reasons it is being developed to DO-178C DAL C.

Other software elements are DAL D with the exception of the SPR Maintenance application and the RT-TipTrak which are DAL E. Originally the functionality of the IFT VMU and IFT Commander were intended to be combined within a single unit. However, the requirement for software certification to DAL C within the SPR™ control functions, lead to the functionalities of the system being segregated into two hardware units, the IFT Commander for SPR™ Control, and the IFT VMU for interface to the tail rotor sensors, blade tracking camera and to facilitate vibration data logging functions.

CONCLUSIONS

The IFT Flight Trial system successfully demonstrated the capability and benefits that can be achieved through In-Flight Tuning. These include:

- 1R vibration levels reduced by approximately 75% over the compromise solution resulting from traditional RT&B
- Compensation for slowly increasing vibration due to factors such as water ingress and blade erosion (ref 3)
- Reduced RT&B flight checks and the ability to cope with large blade dissimilarities as demonstrated in the blind blade change increases aircraft availability

Throughout the flight test campaign the system operated flawlessly and the design proved to be both reliable and robust.

All key features of the IFT Flight Trial System are being developed into a production system for the CH53. A number of improvements over the Flight Trial system have been incorporated into the production hardware/software. In particular:

- Pre-set adjustment of the SPR™ is now electrically driven from the IFT Hand Held Unit
- The IFT Crew Display is a fully featured cockpit display with Night Vision compliance and a textual display to allow maintenance information to be displayed.
- The IFT Commander is being developed to Design Assurance Level C

ACKNOWLEDGMENTS

The flight test program has been carried out under the contract "Development, Fabrication, Qualification and Flight Testing of an IFT System" awarded by the Federal Office of Bundeswehr Equipment, Information Technology and In-Service Support (BAAINBw). The authors gratefully acknowledge the significant support of the flight crews, ground support personnel and the certifying staff of the Bundeswehr Technical and Airworthiness Center for Aircraft (WTD 61) in Manching during preparation and execution of the IFT flight trials. The current production system activities are also carried out under an awarded contract of the Federal Office of Bundeswehr Equipment, Information Technology and In-Service Support (BAAINBw) called "IFT Productionalisation and Type Certification for WaSys CH-53".

Author contact:

Dr. Dominic Southgate domnic.southgate@helitune.com

Dr. Richard Hunt richard.hunt@helitune.com

Douglas Graham douglas.graham@helitune.com

Dr. Daniel Fuerst daniel.fuerst@zf.com

Dr. Uwe T. P. Arnold uwe.arnold@zf.com

REFERENCES

[1]Norman, Thomas R. et al., "Full-Scale Wind Tunnel Test of a UH-60 Individual Blade Control System for Performance Improvement and Vibration, Loads, and Noise Control", 65th Annual Forum of the American Helicopter Society, Grapevine, Texas, May 2009.

[2]Fuerst, D., Arnold, U.T.P. and Graham, D., "In-Flight Tuning: Wind Tunnel Test Results and Flight Test Preparation", 67th Annual Forum of the American Helicopter Society, Virginia Beach, VA, May 2011.

[3]Arnold, U.T.P. and Fuerst, D. "In-Flight Tuning System for the CH-53G Helicopter", 41st European Rotorcraft Forum 2015.

[4]Arnold, U.T.P., Fuerst, D., Hartmann, S., Hausberg, A., Flight Testing of an In-Flight Tuning System on a CH-53G Helicopter, 70th Annual Forum of the American Helicopter Society, Montréal, Québec, Canada, May 2014

Generation of a Tachometer Signal from a Smart Vibration Sensor

Eric Bechhoefer
System Engineer/CEO
GPMS Inc.
Cornwall, VT, USA

ABSTRACT

Because of the bandwidth limitation of the engine controller, all helicopter gearboxes change speed over time. This change in speed necessitates the resampling of the data, based on a tachometer signal, to facilitate shaft, gear and bearing analysis of the rotating equipment. Without resampling, the quality of vibration analysis is degraded and many mechanical faults would be missed. Further, interfacing with existing tachometer can be both expensive, and in some cases change certification requirements. The ability for a smart sensor to acquire vibration data, extract the shaft speed from the vibration data, and then process the data allows vibration based fault detection capability at a lower cost, weight and reduced installation complexity than previously possible. Reducing cost, weight and installation complexity will expand the business case for condition monitoring, improving safety and reliability in industrial and transportation systems. This paper demonstrates a two-step process to recover a tachometer signal from vibration data that is of higher quality than raw tachometer data. Statistics are generated from known fault cases to demonstrate the effectiveness.

INTRODUCTION

R.M. Stewart (Ref. 1) introduced the time synchronous average (TSA) into gearbox analysis, and revolutionized condition based maintenance on helicopters. His analysis, such as FM0/1/2/3/4, the "difference signal", the envelope analysis, etc., are used by most, if no all, Health and Usage Monitoring Systems currently available. Stewart's design is heavily dependent on a tachometer input to allow resampling of vibration data for the analysis. Further, Stewart recognized the importance of tachometer jitter, noting that there was a limit to the number of revolutions to take on the TSA (e.g. at some point, the TSA was not consistent).

The need to interface with a tachometer adds cost and complexity for every application. Further, there may be cases, such as glandless pumps (Boiler Circulator Pumps), where due to heat and pressure it is impractical or infeasible to install a tachometer sensor. In other situation, such as monitoring gas turbine engine, interfacing with the existing tachometer for the power turbine or compressor turbine, may change certification requirements (adding cost) and again increase system cost and weight. These are just a few examples where adding a tachometer is impractical. Hence, the ability to derive a tachometer signal from vibration can reduce both cost and weight of a system, and may facilitate new application were it was impossible to integrate a tachometer.

For helicopter applications, weight and cost of HUMS integration can be reduced through the use of a bused, smart system (Ref 2). The use of a data bus greatly reduces interconnects, while smart sensor makes for a more scalable system. The smart vibration sensor also allows a path to locally processing vibration data to reconstruct a tachometer signal on the sensor, so that the TSA and other tachometer dependent analysis can be performed in situ.

TACHOMETER SIGNAL FROM VIBRATION

The ability for a smart sensor to acquire vibration data, extract the shaft speed from the vibration data, and then process the data allows vibration based fault detection capability at a lower cost and weight. This also results in reduced installation complexity over traditional tachometer based system. Reducing cost, weight and installation complexity will expand the business case for condition monitoring, improving safety and reliability in light helicopters.

Cost is an important driver in the decision for operators to implement a HUMS. While it is well recognized that HUMS improves operational availability and reduces maintenance cost (through an installed Rotor Track and Balance function, opportunistic maintenance, exceedance monitoring, etc.) the decision to install a HUMS for many operators is driven by cost. Removing a tachometer sensor and interface by adding this function to a smart sensor to improve the value proposition of a system to a customer.

The Need for a Tachometer

Spectral content of vibration is measured using the Fast Fourier Transform (FFT). The FFT is used in vibration-based diagnostics to: determine the magnitude and phase of component's vibration (such as shafts, gears or bearings), which can be indicative or wear and failure. Additionally, many common vibration analyses (Residual or Difference Analysis, Narrowband analysis (Ref 3, 4, 5)) use the FFT for ideal filtering of the signal, or to perform a Hilbert transform of the signal (Amplitude and Frequency Analysis).

The Fourier transform (and FFT) assumes that the signal under analysis is infinite in time. This assumption fails for real signals, but can be mitigated by the use of a window function (such as a Hamming or Hanning window). The other assumption in using the FFT is that the signal is stationary. Stationarity implies that the conditions of the signal do not change, e.g. the shaft rate is constant. This assumption in practice is violated: rotating machines have variation in the their shaft rate. This change in rotational rate is due to changing load and the limits of the feedback control bandwidth.

The failure of stationarity results in "spectral smearing" of energy associated with a shaft. This in turn results in not measuring, accurately, the energy associated with a particular fault frequency. In order to improve the performance of analysis using the FFT, Time Synchronous Averaging (the TSA, for shaft/gear analysis) and Time Synchronous Resampling (TSR) have been developed (Figure 1).

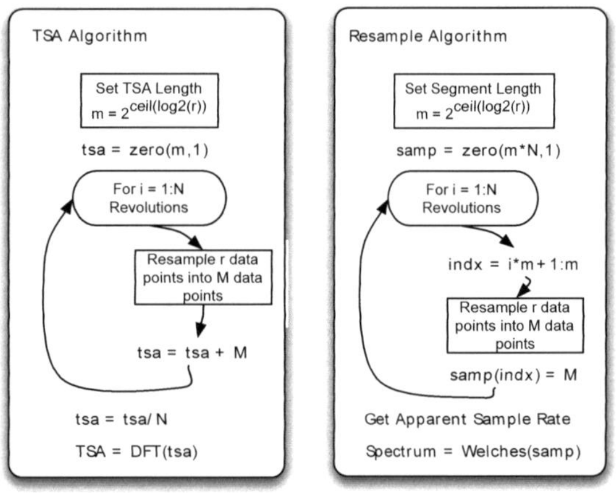

Figure 1 Resampling Algorithms for Vibration Based Diagnostics

Resampling Algorithms

The TSA resamples the vibration associated with a shaft or gear, in the spatial domain, such that vibration associated with each shaft order, in the Fourier domain, represent one frequency bin. For example, consider a system in which the shaft rate is such that for a given vibration sample rate, the acquisition system on average collects 800 samples per revolution. The TSA would resample the 800 samples to the 1024 data points. The value 1024 is the next highest radix 2 value. Radix 2 values are used because the simplest implementation of the FFT is based on powers or 2, e.g radix 2 values.

Consider that if the load on the shaft reduces, the shaft rate increases, and the measured vibration to 780 samples. It takes less time for the shaft to make one revolution, hence the fewer sample rate. The 780 samples are resampled to 1024 points. At some future time, say that the load on the shaft may increase, slowing the shaft, resulting in measuring 820 samples, again the data is resample to 1024 points. For every revolution, the resampled data is summed point by point. After n revolutions, each of the 1024 points of is divided by n, essentially time synchronously averaging the vibration data

Assume in this case that there is a gear with 37-tooth gear on the shaft. The gear mesh energy of the gear would be a frequency of 37 x the shaft rate. In the Fourier domain, energy associated with the shaft rate would be in bin 2, and the gear mesh energy would be bin 38, and the second harmonic of that gear would be in bin 75 (37 x 2 + 1, bin 1 is the DC energy). The TSA also reduces non-synchronous vibration by 1/sqrt(n), where n is the total number of shaft revolutions which where used to construct the TSA. In this way, the TSA corrects for variation in the shaft rate and improves signal to noise.

The TSR similarly resamples (e.g. up samples) the vibration to correct variation in shaft speed. The apparent sample rate is the ratio of the total resampled time domain, vibration data set length divided by original data set length, multiplied by the original sample rate.

Both the TSA and TSR use a tachometer signal to calculate the time over which a shaft completes one revolution. The time taken for any shaft to complete a rotation can be calculated even if the tachometer is not associated with a given shaft. This is done by taking into account the shaft ratio from the shaft with the tachometer, to the shaft under analysis, then interpolating based on that tachometer signal.

The Tachometer Signal

In implementation, the tachometer signal is the rising edge of a voltage trigger from the passing of a shaft key phasor (e.g. a stationary point of the shaft). The tachometer signal is then converted to time. This time is accrued for each pass of the key phasor. In an architecture where the tachometer signal is measured using an analog to digital converter (ADC), the resolution in time of the rising edge is 1 over the sample rate of the ADC. For condition monitoring purposes, the sample rate for a high-speed shaft could be 100,000 samples per second. In another architecture, the tachometer signal inputs into a voltage comparator. When the tachometer signal crosses zero (or some low voltage offset), the comparator goes high. The output of the comparator is monitored by the microcontroller using a general-purpose input/output (GPIO) pin. When the microcontroller senses the GPIO pin going high, it records the time. The resolution of time on the microcontroller is typically much higher than an ADC. For example, in a system using a 12 MHz clock, the microcontroller might run at 96 MHz, but the counter for time in the microcontroller would run at 48 MHz. The tachometer resolution in time would then be 2.0822e-8 seconds, or 480 x greater resolution than the ADC architecture.

Vibration signals from rotating equipment are sinusoidal and, by definition, synchronous with the shaft. However, the nature of vibration makes it impossible to use the vibration signal without significant signal process. Measured vibration is the superposition (e.g. addition) of many signals in time domain. For example, consider a simple gearbox with an input shaft, and output shaft and a gear pair. The input shaft turns at 30 Hz, and has a 32 tooth gear, the output shaft has a 82 tooth gear with a speed of 11.707 Hz. The gear mesh frequency is 30 * 32 = 960 Hz. It is likely that the gear mesh frequency will have side bands as a result of any shaft imbalance being modulated onto the gear mesh. This can be proved using the trigonometric identity:

$$cos(a) * cos(b) = \frac{1}{2}[cos(a+b) + cos(a-b)]$$

Where in this example, cos(a) is 960 Hz, and cos (b) is 30 and/or 11.707 Hz shaft. Additionally, if the shaft is bent or bowed, there will be a 2x shaft vibration component. Other manufacturing defects such as the gear not being mounted perpendicular to the shaft, or not centering the shaft on the gear (e.g. eccentricity) adds additional tones.

TACHOMETER FROM VIBRATION

Bonnardot (Ref. 6) in 2005 describes a method of using an acceleration signal of a gearbox to perform the TSA. This was further developed by Combet (Ref. 7) in 2007. These methods call for band pass filtering of the signal around a gear mesh, then in (Ref. 6) taking the Hilbert transform to estimate phase, and using phase to estimate zero crossing time. In (Ref. 7), the signal was band pass filtered then demodulated, with the time resulting time domain signal used for the zero crossing.

Both solutions are sub optimal, due to the nature of using the FIR (Finite Impulse Response) band pass filter. Additionally, because modeling errors are present even after band pass filtering, the tachometer signal is corrupted by jitter. In this paper, we present a two-step method to reconstruct a tachometer signal from vibration. Step one is to ideally band pass filter and create an analytic signal in one functional procedure. Step two uses a jitter reduction model to remove noise (jitter) from the reconstructed tachometer signal not associated with changes in the shaft rate. While this system could be implement on any computer, it is ideally suited for implementation in a smart sensor, where cost and weight can be better addressed by lower cost condition monitoring system.

The pseudo code to recover a tachometer signal from vibration is:

Define the Sample rate = sr. The number of data points of vibration data, $n = sr$ x acquisition length in seconds, then:

1. Calculate the next larger radix-2 length for the FFT. nRadix = 2^ceil($\log_2(n)$)
2. Calculate the low and high bandwidth index (*bwlow, bwhigh*), which are centered are a know gear mesh
3. Take the zero padded FFT of the vibration data
4. Zero the FFT from zero to *bwlow*, and from *bwhigh* to nRadix
5. Take the inverse FFT
6. Calculate the unwrapped argument of the signal from 1 to n time series
7. Normalize the time series of radians by the number of teeth of the gear (assuming 1st harmonics)
8. Interpolate the number of indexes for every 2π radians
9. Normalized to tachometer zero crossing times by sr.

A band pass filter is the convolution of a low pass filter with a high pass filter. These filters are implemented as Finite Impulse Response (FIR) filters to improve their stability. Unfortunately, even rather large tap filters have poor filter response. Consider a case of a wind turbine gearbox, with an aforementioned approximately 29 Hz shaft and a 32 tooth gear, giving 928 Hz. The bandwidth of the filter is set from 910 Hz to 950 Hz, to exclude the 30 Hz of the high-speed shaft (e.g. 29 * 32 -29 = 899 Hz and e.g. 29 * 32 + 29 = 957Hz). The filter response for this case, using a 120 tap FIR filter is seen in Figure 2.

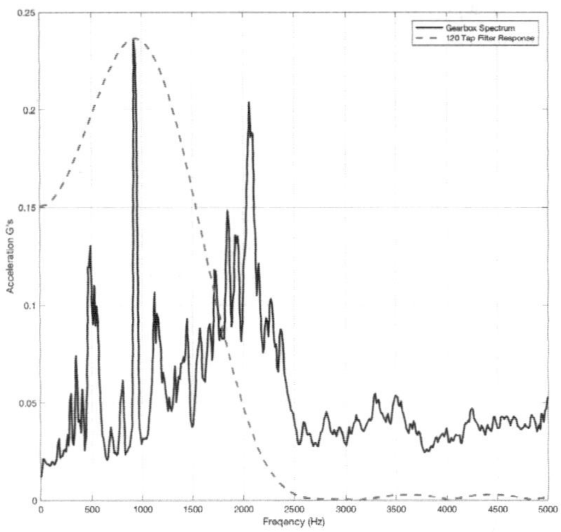

Figure 2 Spectrum from a wind turbine high-speed shaft, and a 120 tap FIR band pass filter

Note that the bandwidth (50% power, 3dB) of this filter is 1670 Hz. The filter does not reject the spectral content at

500 Hz, or at 1121 Hz. These additional tones distort the desired analytic signal, reducing the quality of the resulting tachometer signal.

Consider a process in which the developed analytic signal uses an ideal filter. The analytic signal is defined for real valued signal *s(t)* as Eq 1.:

$$S(f) = F\{s(t)\} \qquad (1.)$$

Where *F* is the Fast Fourier Transform, then:

$$S_a(f) = S(f), \qquad f = 0$$

$$S_a(f) = 2S(f), \qquad f > 0$$

$$S_a(f) = 0, \qquad f < 0$$

$$s_a(t) = F^{-1}(S_a(f))$$

where *S(f)* is as noted the Fourier transform of *s(t)*.

For a signal which is sampled at 97656 samples per second, for six seconds, the total length of *s(t)* is *n*, 585936 data points. As noted, there are advantages to using radix 2 lengths for the Fast Fourier Transform (FFT). By zero padding the FFT to next larger radix two value, 2^20 or 1048576, the index representing the cutoff frequency for the band pass frequencies are: *bwlow* = 910 Hz / 97656 x 1048576 = 9771, and *bwhigh* = 960 / 97656 * 1048576 = 10308. Then one can define the band pass analytic signal in Eq 2:

$$S_a(f) = 2S(f), \qquad bwlow \leq f \leq bwhigh \qquad (2)$$

$$S_a(f) = 0, \qquad f < bwlow, f > bwhigh$$

$$s_a(t) = F^{-1}(S_a(f))$$

There is no need to multiply by 2, as the argument (e.g. angle) which is of interest, is the arctangent ratio of the imaginary parts of $s_a(t)$ and the real parts of of $s_a(t)$. The idealized band pass function reject all signals not associated with the desired pass band. Figure 3 is a zoomed view of the spectrum in Figure 2., comparing the pass band of the idealized filtered realized using Eq 2.

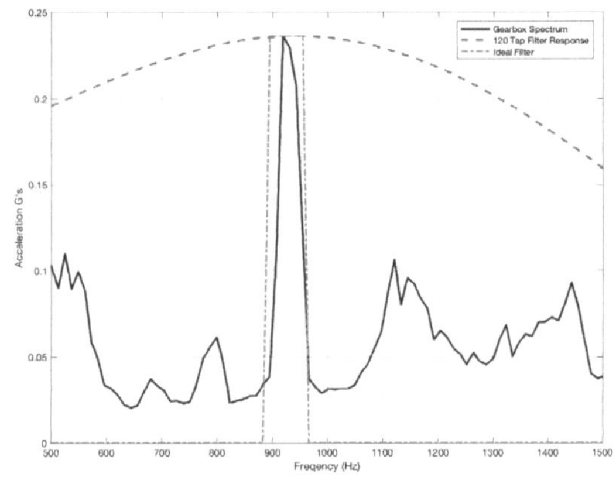

Figure 3 FIR band pass filter vs. the idealized filtered realized using Eq 2.

The FIR filter does not reject the 500 Hz and 1150 Hz tone. The idealized filter captures only the signal associated with the desired gear mesh tone. This allows for higher signal to noise ratio and improved reconstruction of the tachometer signal from vibration. This tachometer signal is recovered from the arctangent of the analytic signal.

It is important to note that the arctangent function returns radians between 0 to π and -π to 0. But the evolution of the angle represents the incremental increase in phase for each sample in time. For example, the phase of the analytic signal for three cycles is: 2π x 3 or 6π. The arctangent of that signal will be -π to π for 3 cycles. The result of the arctangent must be unwrapped to capture these increases in phase vs. time. Unwrapping of the phase angle requires keeping track of the previous angle and current angle. The current angle is added to the previous angle, except when the returned arctangent goes from π to -π. In this case, π is added to the returned value to correct for the case when the returned value is between -π to 0.

After unwrapping the phase angle, the units are in radians per sample. While the FFT and inverse FFT operated on the radix 2 length (in this case 1048576), the argument and phase angle computation are performed only on the original sample length, *n*. Note that this time series of radians is for the gear mesh. To converts to radians per revolution of the shaft, the time series is divided by the number of teeth in the gear: in this example, 32.

However, it may be that the strongest gear mesh tone is the 2nd or 3rd harmonic, in which case the pass band is adjusted accordingly, and the time series of radian angle is divided by 2 x number of teeth for the 2nd harmonic, etc.

The resulting time series represents the radian angle of the shaft, were each index advances the angle in time by *dt*, or 1/sample rate. Every 2π radians represents one shaft revolution. Because one is interested in the time, exactly every 2π, a form of interpolation is needed. For example, consider that the index just prior to 2π is 6.282780795474 (or 0.0004 less than 2π) at array index 3395, while at index 3396, the radian value is 6.284629142378, or 0.0014 greater than 2π. It is then a simple matter to interpolate between the index 3395 and 3396 for a radian value of 2π. In this case, the interpolated value is: 3395.21885053316.

This interpolation gives the number of index for each revolution. Note that this estimate of the tachometer zero cross signal is corrupted by noise. Combert (Ref. 7) reports that the phase error standard deviation is related to the local signal to noise at the mesh harmonic *k* as:

$$\sigma[\delta] = \frac{1}{\sqrt{2}} 10^{-SNR/20}$$

Typical measured SNR are 6 to 8 dB. This suggests that the standard deviation of the phase error would be 6 to 10 degrees. This phase is zero mean, but as it is non-zero: it will add jitter to the reconstructed tachometer signal.

Controlling Jitter

In "Improving Gear Fault Detection by Reducing Tachometer Jitter" (Ref 8), it was shown that tachometer jitter contained a low a low frequency component associated with the engine control unit, and random, higher frequency components. Bechhoefer et. al. improved gear fault detection by using a zero phase, low order IIR, backward/forward filtering. As noted previously, both FIR and IIR filters bandwidth is defined by the 3dB reduction in signal energy. The filter does not remove all signals above the bandwidth, and in fact reduces some signal energy below cutoff (up to 50%).

The idealized filter using the FFT processing is zero phase as well. The improve analysis proceed similar to the forward/backward process:

1. Take the pseudo derivative of the tachometer signal
2. Calculate the the radix-2 length of the pseudo derivative signal of length n
3. Zero pad the array from n to the radix-2 length
4. Calculate the bandwidth index of the FFT
5. Idx = floor(bandwidth * radix-2 length / 2);
6. Bandwidth is a normalized value, typically 0.12
7. Take the real FFT of the zero padded derivative signal
8. Set the real and imaginary parts of the FFT from Idx to the radix-2 length
9. Take the inverse real FFT.
10. Reconstruct the tachometer signal by taking the pseudo integral of the signal

Figure 4 demonstrates the effectiveness of the ideal low pass filter in removing jitter from the tachometer signal. The tachometer was from a wind turbine with 8 targets per revolution. The tachometer signal was resample and the spectrum taken.

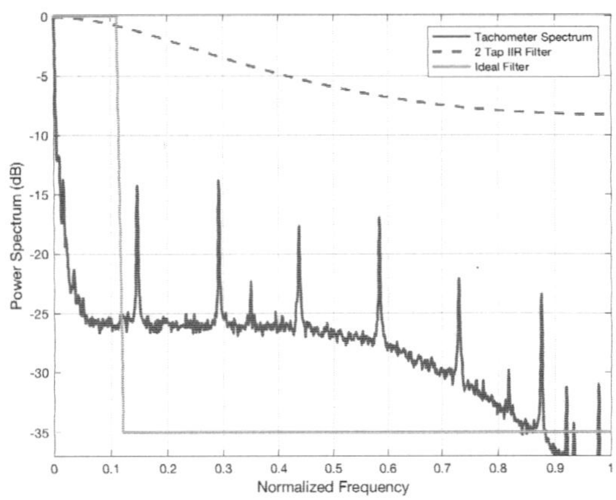

Figure 4 Tachometer pseudo derivative spectrum compared to the IIR and ideal low pass filter.

Note that the tachometer spectrum is primarily DC, with peaks associated with the unequal spacing of the targets. The idealized filter retains more low frequency response than the IIR filter, and effectively removes all spectral power (jitter) above the cutoff frequency.

CASE 1: HIGH SPEED PINION ON A WIND TURBINE

This machine was found, through vibration based diagnostics, to have a cracked tooth on the high speed pinion. The acquisition length was six seconds, sampled at 97656 samples per second. The high-speed shaft is approximately 30 Hz.

This system was equipped with a tachometer, which is used as a reference to compare against the vibration based tachometer signal. The installed tachometer was a Hall sensor, using the GPIO architecture to capture zero crossing time. The tachometer target had eight pulses per revolution, but time was calculated using each 8th timing mark (e.g. one revolution). Figure 5 compares shaft speed derived from the Hall sensor base tachometer signal, vs. the vibration based tachometer signal, vs. the final processing use the new jitter reduction technique. Wind turbine analysis is particularly difficult in that there is a high degree of variability in the torque.

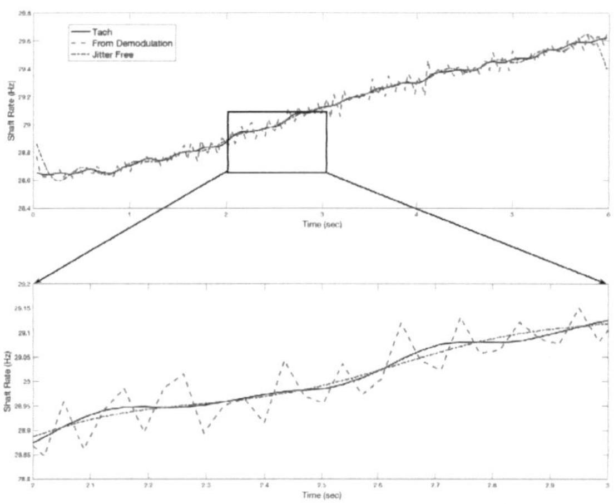

Figure 5 Shaft Rate from Tachometer signal in comparison with jitter free tachometer from vibration

The bottom graph is a zoomed from the 2nd to 3rd second on the acquisition. In general, after removing jitter using the new process, the vibration based tachometer signal is not significantly different from the tachometer derived from the Hall sensor. Figure 6 depicts the TSA and the spectrum of the analysis.

The difference between the TSA is effectively only phase angle. From the spectrum, we can calculate the signal to noise ratio to be approximately 6. Statistics derived from the TSA are given in table 1.

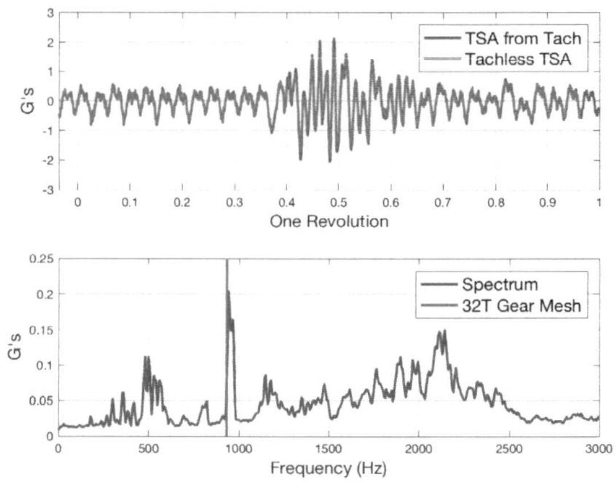

Figure 6 Comparison of the TSA derived from a tachometer and vibration sensor

Table 1 Tach vs. Tachless for Standard Vibe Statistics

Analysis	Tach	Tach from Vibe
SO1	0.0100 g	0.0104 g
SO2	0.0013 g	0.0016 g
SO3	0.0019 g	0.0018 g
TSA RMS	0.5091 g	0.4828 g
TSA P2P	2.0887 g	1.8430 g
FM0	4.278	4.00
AM RMS	0.100 g	0.099 g
AM Kurtosis	4.242	4.217
FM RMS	0.428 radians	0.426 radians
FM Kurtosis	4.995	4.844

Clearly, the gear fault was detectable with either a tachometer, or via vibration method.

CASE 2: TAIL ROTOR INTERMEDIATE GEARBOX

This was taken from public domain data set, donated by NAVAIR AIR 4.4.2 (Ref 9.). The sample rate was 100000 samples per second, using the ADC architecture for recording zero crossing data. The tachometer sensor was a variable reluctance speed sensor, using a 22-tooth gear as a target. The shaft rate for the target gear was approximately 3000 rpm (e.g. 500 Hz), with a ratio from this shaft to the tail rotor drive shaft of 7.3:1. As seen in Figure 8, the tachometer signal is noisy (high jitter), with a mean shaft rate of 68.5 Hz. The zoomed view shows the tachometer from vibration and with less jitter than the VR sensor tachometer signal, while the jitter free tachometer signal is smooth, capturing a slowly evolving change in shaft rate.

The SNR for this example is approximately 12 dB, due to the 65g mesh frequency (Figure 9). The resulting TSA, aside from a difference is phase, is indistinguishable from the TSA using a tachometer from the VR sensor.

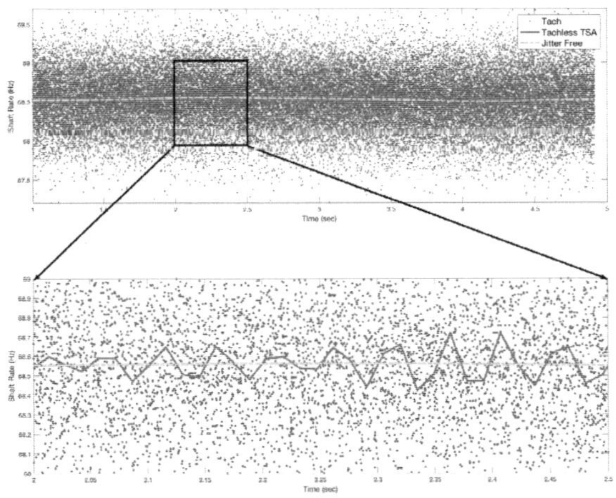

Figure 7 VR Tachometer vs. tach and jitter free tach from vibration

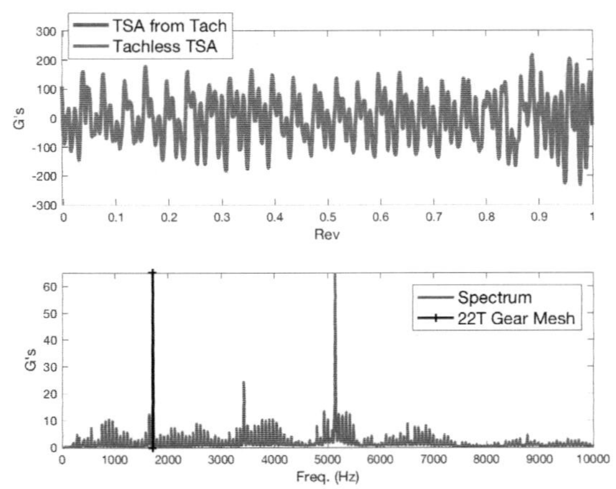

Figure 8 Comparison of the TSA derived from a tachometer vs. vibration sensor

Statistics derived from the TSA for the tail rotor input gear are given in table 2:

Table 2 Tach vs. Tachless for Standard Vibe Statistics

Analysis	Tach	Tach from Vibe
SO1	0.043g	0.043g
SO2	0.283 g	0.282 g
SO3	1.855 g	1.85g
TSA RMS	81.646 g	81.591g
TSA P2P	222.77g	223.02g
FM0	16.94	15.38
AM RMS	17.06g	17.06g
AM Kurtosis	4.126	4.126
FM RMS	3.037 radians	3.048 radians
FM Kurtosis	2.46	2.45

CONCLUSIONS

A method to reconstruct the tachometer signal from vibration was demonstrated using a two-step process. The first step uses an idealized band pass filter implemented with a complex FFT to create an analytic signal, which is needed to derive the tachometer signal for the shaft under analysis. The second step uses another idealized low pass filter on the pseudo derivative of the reconstructed tachometer signal to remove jitter. This processing is run on a smart vibration sensor, which facilitates improved vibration analysis on rotating equipment where in the past the addition of a tachometer would be prohibitive due to cost, weight, certification requirements or being physically impractical.

The quality of the resulting analysis is compares favorable with traditional tachometers on real world fault data. In fact, the tachometer from vibration signal may have less noise, resulting in improved fault detection, than traditional tachometer sensor based system.

Author contact: Eric Bechhoefer eric@gpms-vt.com

REFERENCES

[1]Stewart, R.M., ìSome Useful Data Analysis Techniques for Gearbox Diagnostics,î *Institute of Sound and Vibraiton Research,* Proceeding of Meeting on Application of Time Series Analysis, Sept. 1977

[2]Bechhoefer, E., Augustin, M, Kingley, M., ìArchitecture for a Light Helicopter HUMSî, AHS Forum 68, May 1, 2012.

[3] Bechhoefer, E., Van Hecke, B., He, D., *Processing for Improved Spectral Analysis*, PHM Society Annual Forum, 2013.

[4]M. Lebold, K. McClintic, R. Campbell, C. Byington, K. Maynard, "Review of Vibration Analysis Methods for Gearbox Diagnostics and Prognostics", Proceedings of the 54th Meeting of the Society for Machinery Failure Prevention Technology, 2000.

[5]S.A. McInerny, Q, Sun, B. Hardman, "Public Domain Algorithms for Bearing and Gear Fault Detection", Proceedings of the 11th International Conference on Structure Borne Sound and Vibration, 2004.

[6]Bonnardot, F., El Badaoui, M., Randal, R. B., Daniere, J., Guillet, J., "Use of the acceleration signal of a gearbox in order to perform angular resampling (with limited speed fluctuations)", Mechanical Systems and Signal Processing 19 (2005) 766-785.

[7]Combet, F., Gelman, L., "An automated methodology for performing time synchronous averaging of a gearbox signal without speed sensor", Mechanical Systems and Signal Processing 21 (2007) 2590-2606.

[8]Bechhoefer, E., He, D., ìImproving Gear Fault Detection by Reducing Tachometer Jitterî, AHS Forum 72, May 1, 2014.

[9]NAVAIR AIR 4.4.2, Helicopter Drivetrain Vibration Condition Monitoring Algorithms, http://web.archive.org/web/20150514044141/http://qsun.eng.ua.edu:80/cpw_web_new/index.htm

Wireless Sensor Suite for Electrical System Diagnosis and Prognosis

Michael W. Usrey
R. Ray Depew
Laurel Frediani
Brian Schaible
Dale Schoonover
Lewis Wright
Alex Brand
Sporian Microsystems, Inc.

Patrick Kalgren
Singularity-Intelligence Amplified LLC

Al Duke
UTC Aerospace Systems (UTAS)

R. Steve McKown
Titanium Mirror, Inc.

Chris Lyman
US Army Research Development and Engineering Command, Aviation Development Directorate

ABSTRACT

The authors have developed a wireless sensor suite for rotorcraft Generator Control Unit (GCU) and Main Power Relay (MPR) health monitoring. This sensor suite monitors for changes in component characteristics, temperature, vibration extremes, voltage surges, and other factors that will indicate the unit is close to the end of its service life. The sensor suite logs event/triggered or time sequenced/snapshot "smart data" for prognostic and diagnostic estimations, then wirelessly transmits the logged data to maintenance personnel and/or a Health Monitoring and Usage System (HUMS) mounted on the rotorcraft. The system assists in diagnosis and corrective maintenance by capturing data at the time of a fault and aiding in its visualization. This system also assists with Condition Based Maintenance (CBM) by calculating prognostic signatures that indicate to maintenance personnel when failure of select electrical system components is imminent.

NOMENCLATURE

CBM	=	Condition Based Maintenance
COTS	=	Common Off-the-Shelf
DoD	=	Department of Defense
DTU	=	Data Transfer Unit
ESA	=	Electrical Signature Analysis
FCS	=	Frame Check Sequence
FD/FI	=	Fault Detection/Fault Isolation
FIFO	=	First In, First Out
GCU	=	Generator Control Unit
GSE	=	Ground Support Equipment
HIL	=	Hardware in the Loop
HUMS	=	Health and Usage Monitoring System
IVHMU	=	Integrated Vehicle Health Monitoring Unit
MPR	=	Main Power Relay
OSA-CBM	=	Open Systems Architecture-CBM
PHM	=	Prognostics and Health Management
PSEF	=	Propulsion Systems Evaluation Facility
RCM	=	Reliability Centered Maintenance
RMS	=	Root Mean Square
SD/HA	=	State Detection/Health Assessment
SIL	=	Systems Integration Lab
VT	=	Visualization Tool
WBI	=	Wireless Bridge Interface

INTRODUCTION

There is a need within the Department of Defense (DoD) to reduce maintenance costs. Currently, the military replaces components on mission critical vehicles (e.g. rotorcraft) on a time-in-service basis. However, this maintenance procedure typically causes components to be removed that still have service life remaining. The DoD is moving away from a time-in-service replacement of components to Reliability Centered Maintenance (RCM). RCM is an optimal deployment of corrective, time-based and Condition Based Maintenance (CBM) (Ref 1). Electrical and electronic components are typically replaced when a malfunction occurs, or as some like to say, "on-condition." The generator control unit (GCU) suffers a high occurrence of incorrect replacement, as diagnostic ambiguity typically exists which precludes accurate on-aircraft fault isolation. The GCU may be part of an ambiguity group and it may be expedient to replace it in an attempt to reduce the group size, affecting a final diagnosis. Unnecessary replacements result in a substantial waste of time, money, and resources. Thus there is a need to develop a system that can actively and in real time provide information on the condition of the GCU, predict when the component is going to fail and aid in the diagnosis and repair of failed units.

Distribution Statement A: Approved for public release; distribution unlimited. Control Number PR3536

This system would have to monitor operating and environmental conditions and then, using changes in sensed characteristics in conjunction with reliability calculations, determine the remaining useful life of a component. Such a system would result in overall improvements in Condition Based Maintenance (CBM+) for GCUs, Main Power Relays (MPRs) and potentially other main power system components. Ultimately, such a system would result in overall better availability of aircraft fleets and reduced operating costs by reducing aircraft down time and unwarranted maintenance procedures.

To address this need, Sporian Microsystems, Inc. is developing a wireless sensor suite for rotorcraft GCU and MPR health monitoring. This PowerIQ™ sensor suite will monitor for changes in operating characteristics (e.g. resistance of relay contacts, voltage surges) and environmental conditions (temperature, vibration) and other factors that will indicate the unit is close to the end of its service life. The sensor system will log periodic and event-driven "smart data" for on-aircraft and ground-based processing supporting prognostic and diagnostic processes, then wirelessly transmit the logged data to maintenance personnel and/or a HUMS mounted on the rotorcraft. The system under development provides prognostic signatures via wired connection to the HUMS and provides an on-aircraft FIFO log that can be accessed wired or wirelessly by maintenance personnel.

SYSTEM OVERVIEW

The PowerIQ system can give the insights needed to provide CBM information on the power generation systems on a fleet of helicopters. The complete system will monitor all the GCUs and MPRs in the rotorcraft by performing real-time smart data collection of the necessary electrical, thermal, and vibrational signals on the GCU, MPR and potentially other electrical system components. The monitoring system will then transmit the collected data to a Wireless Bridge Interface (WBI) module. The system approach is represented schematically in Figure 1.

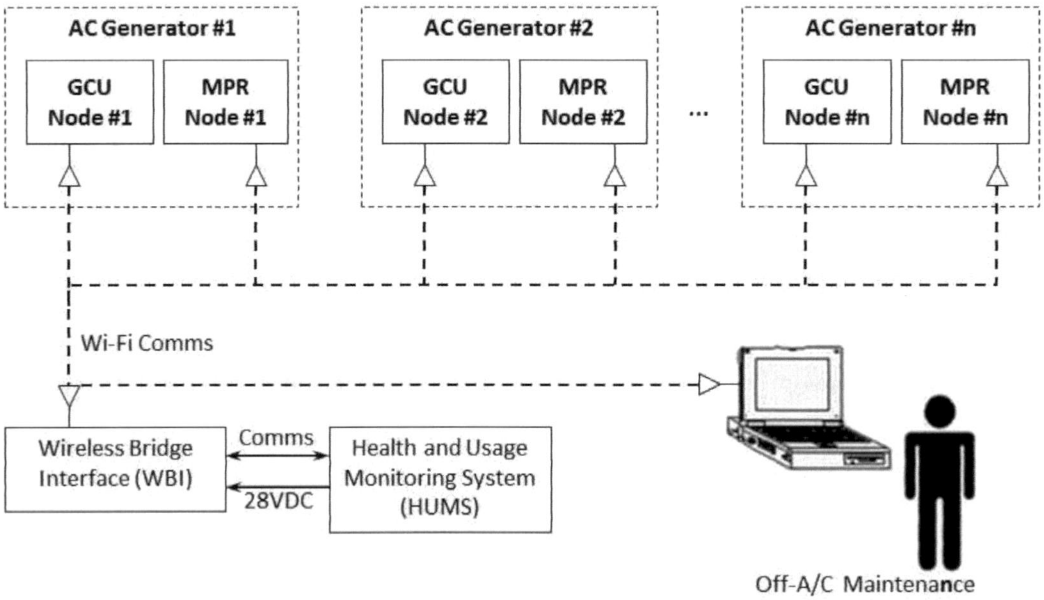

Figure 1: PowerIQ GCU/MPR monitoring system block diagram

The system utilizes separate pass-thru "edge" sensor modules for each GCU and MPR on the aircraft. These electrical system analysis (ESA) nodes comprise a set of high-frequency capture and analysis devices that reveal deep insight into both electrical and mechanical component health in these assets. The system also features a single WBI module to collect, store data, process, and distribute information for the entire aircraft. The WBI can distribute data to on-board systems as well as off-aircraft systems to facilitate prognosis and diagnosis.

Each ESANode™ module features sensors, analog and digital signal conditioning, processing, volatile and non-volatile memory, and communications (**Figure 2**). Wireless communications are well suited to retrofit applications or where weight is a concern. Wired communications can be used in forward-fit applications or where wireless is otherwise undesirable. ESANode functionality could also be incorporated into future "smart" GCU and MPR components.

Distribution Statement A: Approved for public release; distribution unlimited. Control Number PR3536

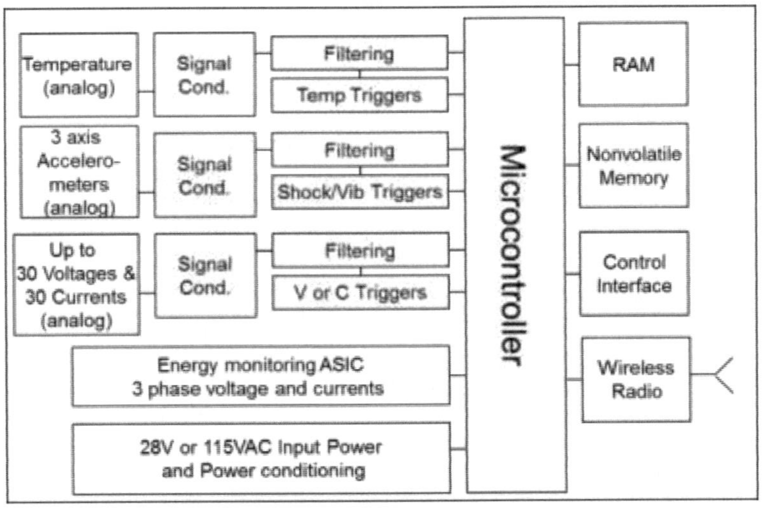

Figure 2: High level Block diagram of GCU/MPR ESANodes

The H-60 Blackhawk was chosen as the first platform for development, demonstration, and possible transition of this technology. Specifically:

- The H-60 fleet is the Army's largest and most utilized helicopter,
- Contracts for new H-60s are being issued from multiple agencies (e.g. Army/Navy and Air Force had issued contracts for over $2.6B for new/modified H-60s in 2014/15),
- Commercial opportunities with S-92 series.

Sikorsky is the manufacturer of the H-60 airframe. The H-60 includes 3 independent power systems: one associated with each main engine and one associated with the auxiliary power unit (APU). UTC Aerospace Systems (UTAS) manufactures the Integrated Vehicle Health Monitoring Unit (IVHMU). Honeywell manufactures the generator and the GCU while TE Hartman manufactures the MPR.

Under the current concept of operations, the ESANodes measure data continuously in flight, capture and store in flight data under specified conditions, and wirelessly communicate to the WBI only while on the ground. Raw diagnostic data are available directly from the WBI to ground crews via wired or wireless links. Prognostic signatures are communicated via wired connection to the IVHMU for aggregation.

DIAGNOSTIC APPROACH AND BENEFITS

ESANodes continuously measure 22 relevant GCU parameters and 26 relevant MPR parameters. Measurement captures are initiated under two scenarios. In the first, the firmware captures and stores periodic "snapshot" data. These data are used primarily for trending, signature recognition, and other prognostic purposes. The prognostic approach and benefits are discussed in the following section.

The module firmware also detects fault conditions to capture and store "triggered" data. Capturing data at the time of the fault is far more cost effective than trying to replicate the fault later. It should be possible to use this data to automate fault trees and other diagnostic methods described in the maintenance manual, though this currently falls outside the scope of this effort. Additional operating modes such as enabling Wi-Fi in-flight could potentially aid in electrical system diagnosis (i.e.: inflight troubleshooting, continuous flight data acquisition for product improvement).

To demonstrate some of the diagnostic potential and to demonstrate retrofit potential in non-HUMS-equipped vehicles, signal visualization software was developed. PowerIQ Visualization Tool (VT) provides a graphical view of the binary data, which has been captured by the GCU ESANodes and MPR ESANodes, and aggregated by the WBI. WBI data are available either wired or wirelessly to VT. VT also has the ability to export the data to a CSV file for use in other applications.

Unfiltered lab data and the VT filtering dialog are shown in Figure 3. Along the bottom of the screen is the timeline. The timeline allows the user to view the various data captures and select which to plot. Up to five different types of data captures can be displayed on the timeline: periodic/snapshot (blue triangle), event/triggered (red triangle) plus up to three different condition indicators per ESANode (red, yellow and blue circles). The timeline can be zoomed and panned to make it easier to see all of the captures in relation to each other. The user selects a capture by clicking on the capture's symbol with the mouse.

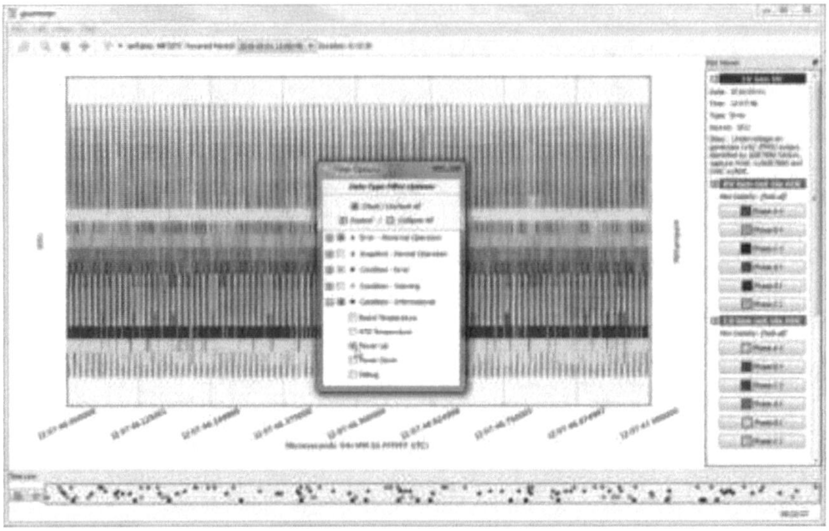

Figure 3: Visualization Tool displaying unfiltered ESANode data and filter dialog

The data display area is below the toolbar in the upper left portion of the screen. Upon selection of a capture, VT plots all of the data in the capture. The data display can be zoomed and panned in order to examine the data more closely.

To the right of the data display, a sidebar called the Plot Viewer shows information about the capture being displayed, and an interactive legend for the data being plotted or filtered. When several parameters are being plotted at once, clicking on the name of a parameter in the legend toggles the plot of that parameter, allowing the user to choose the parameter(s) to display.

VT is written in Python, with a Qt-based graphical user interface, which allows it to run on either a Linux system or a Windows system, as long as the system has Python 2.x installed. An executable (.exe) version is also available, allowing the program to run on a Windows system without Python.

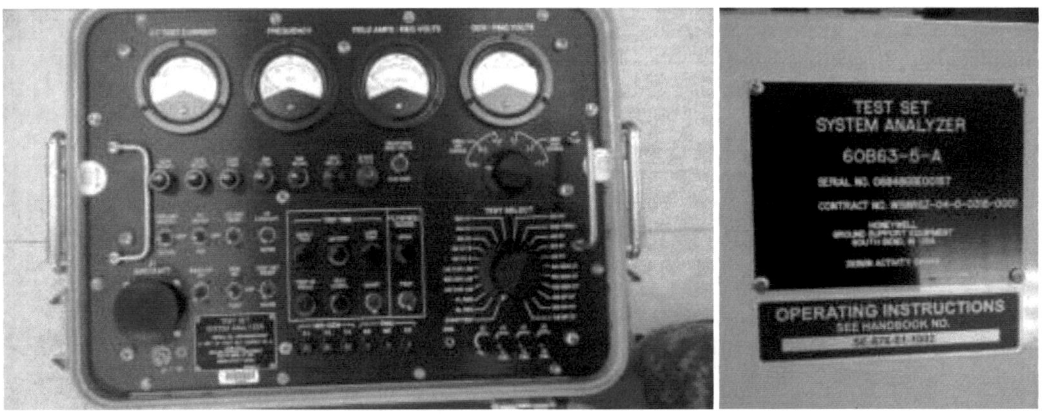

Figure 4: Current Ground Support Equipment - H-60 electrical system Test Set System Analyzer

At its current state of development, VT is a powerful, 'smart' breakout box. Combined with ESANode real-time data acquisition and WBI data aggregation, PowerIQ offers significant technical advantage over using current ground support equipment (GSE - Figure 4) that requires replication of faults after the fact. The system also provides a foundation for heuristic diagnosis and prognosis of electrical systems.

PROGNOSTIC APPROACH AND BENEFITS

CBM recognizes that systems degrade and wear out, allowing anomalous behavior to be used as evidence to track and potentially indict a developing and impending fault. Having access to this history of anomalous behavior provides greater assurance that a true fault, when it develops, will be found and appropriately isolated with less time and expense. Prior work in electrical power generation systems has demonstrated that access to electrical signal data may enable greater fault coverage (such as windings and control and support prognostics) through enhanced detection of incipient fault conditions (Ref 2).

In PowerIQ, prognosis is based on "snapshot" data captured periodically by ESANodes under nominal operating conditions. While initial prognostic algorithms have been developed, the data afforded by the subject monitoring system creates an opportunity to develop even more powerful prognostic tools.

The hierarchy described by the Open Standards Architecture for Condition-Based Maintenance (OSA-CBM) provides a logical starting point for the GCU/MPR fault detection, fault isolation, and prognostics development and implementation (Ref 3). This hierarchy supports the separation of major functions of a condition monitoring system and provides associated standards for communications between these modules.

OSA-CBM specifies seven layers of data and processing including: Data Acquisition, Data Manipulation, State Detection, Health Assessment, Prognostic Assessment, Advisory Generation, and finally Information Presentation. The signal processing algorithms perform the Data Manipulation function. Fault Detection and Fault Isolation (FD/FI) are performed in the State Detection and Health Assessment (SD/HA) blocks respectively. It is noteworthy that condition-based maintenance uses different terminology than standard diagnostics. FD/FI, from the diagnostic perspective, refers to "hard faults," which require immediate action. SD/HA encompasses both traditional FD/FI and adds the possibility of a condition between healthy (new, functional) and faulted (broken, non-functional). Thus, the detection of a state other than new or functional may not always require the generation of an actionable alert, but it can be used to track the changing state (SD) of the monitored system; this tracked state history can be used to improve fault management and enhance diagnostic certainty through a better informed health assessment (HA). From a diagnostics perspective, this gradient between healthy and faulted has been described as "grayscale" health in some standards bodies such as IEEE SCC20 (Ref 4). Generator and electrical power distribution systems straddle a boundary between electronic and mechanical systems, yet diagnostics for these systems typically follow the healthy versus failed paradigm. A useful application of prognostics in this domain is risk assessment rather than remaining useful life prediction. Recognizing the changing risk of failure for a given component, due to measured performance change, can support diagnostic ambiguity reduction.

The purpose of enhanced fault detection is to extract indicators (data features) which react to incipient faults as early in their progression as possible. The challenge to the subject effort is that, while a good background of prior work exists in mechanical and rotating machinery (Ref 5, 6), and some prior work exists in the electro-mechanical (motor) fault detection domain (Ref 7), a smaller background is available to draw from in generator electro-mechanical systems. Table 1 relates features to faults in a rudimentary way.

Table 1. Basic generalization of fault types and potential ambiguities

Generator/GCU Fault		GCU	Generator	Other
Category	Type			
Frequency	Over			Y
	Under			Y
Voltage	Over	Y		Y
	Under	Y		Y
Load	Overload		Y	Y
	Phase Imbalance		Y	Y
Winding Degradation	Phase to Phase		Y	Y
	Phase to Neutral		Y	Y
	Phase to Ground		Y	Y
Bearing/Shaft	Imbalance		Y	Y
	Bearing Defect		Y	
	Shaft Coupling		Y	Y

A primary take-away should be the degree of overlap exhibited between features and fault types. Overlap indicates potential ambiguity when simple feature thresholds are used. This supports the goal of fault detection, which is to indicate that a fault exists, but it is problematic for fault isolation. Thus, fault isolation and health assessment require another level of processing.

Candidates for prognosis include a dominant electronic circuit failure mode of the field exciter in the GCU and electrical failure modes of the generator itself, including generator winding faults, and, provided access to vibration data to support a data fusion process, the future potential for enhanced bearing and shaft coupling coverage. The monitor collects data essential for insight into these failure modes, and while run to failure data is not currently available to facilitate full prognostic model development, the team developed an initial feature set and an offline process for trend tracking and analysis. This offline process is envisioned to enable parametric refinement of physics-based modeling, such as Arrhenius-based aging parameters, as actual failure experience permits. Detecting and monitoring changing trends presents a more immediate opportunity to support enhanced diagnostics and directly supports updated reliability-based risk prediction. Risk analysis and prediction improves diagnostics by providing a condition-based updating of diagnostic candidates for ambiguity group reduction.

CURRENT STATUS

ESANode hardware for H-60 GCU and MPR has been developed via a spiral development method. The final target form factors are shown, respectively, in **Figure 5** and **Figure 6**. The GCU ESANode is connected in-line between the GCU and the cable harness normally connected to the CGU. The MPR ESANode is installed on the top of the MPR and is electrically connected in-line between the MPR and the cable harness normally connected to the MPR. The MPR ESANode includes three current sensors on an external cable assembly. These sensors clamp on to the MPR terminal posts to allow high-bandwidth measurement of current on all three phases of the electrical system.

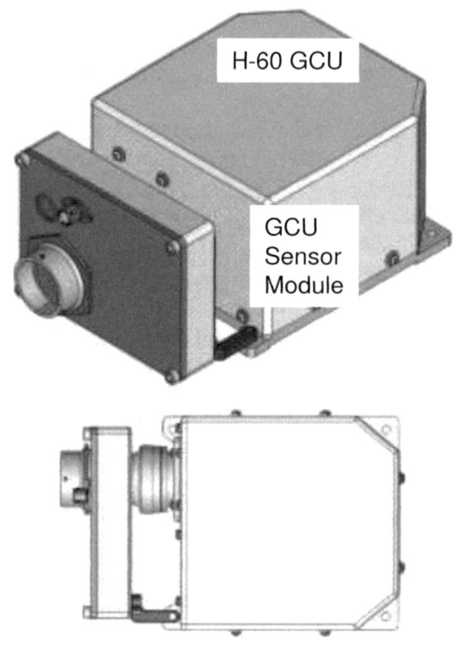

Figure 5: GCU ESANode enclosure in its final form factor, attached to a representative H-60 GCU.

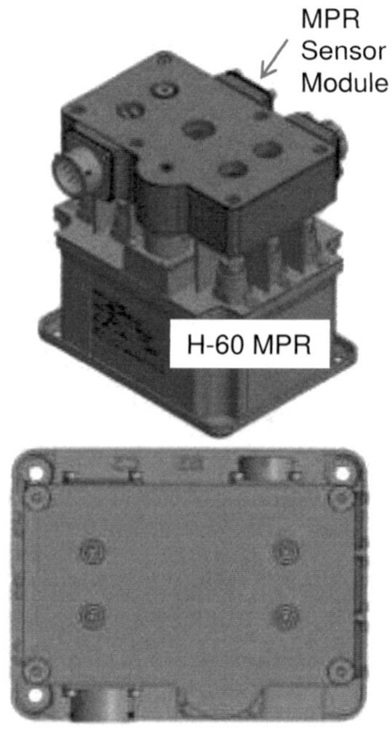

Figure 6: MPR ESANode enclosure in its final form factor, attached to a representative H-60 MPR.

Prototypes in final form factor are shown in **Figure 7**. These edge nodes are currently being subjected to environmental testing. The main goals of the environmental testing are to determine if the devices will be capable of passing a MIL-STD-810 testing suite, and to guide the redesign for next-generation versions of the devices.

MIL-STD-810-oriented temperature testing is currently underway. The characterization is comprised of high and low temperature operation and electronic component performance tests, as well as thermal cycling tests. Complicating the development of the characterization plans was uncertainty of the details of the operational environments for the ESANodes and WBI in various UH-60 models. The possible range of H-60 deployment locations leaves the exact temperature ranges for thermal testing unknown. To circumvent these details, the devices are being tested at the largest operational range possible: high-temperature testing to +85° C and low-temperature testing to -40° C. Performance aspects such as communications, Wi-Fi, and processor operations are being assessed. As part of the thermal testing, at both high and low temperature, each device is unpowered and allowed to soak at the chamber temperature, then repowered to determine if this causes any boot failures. This is done to account for temperature differences between the chamber and the PCBs due to power dissipated by the components on the PCBs.

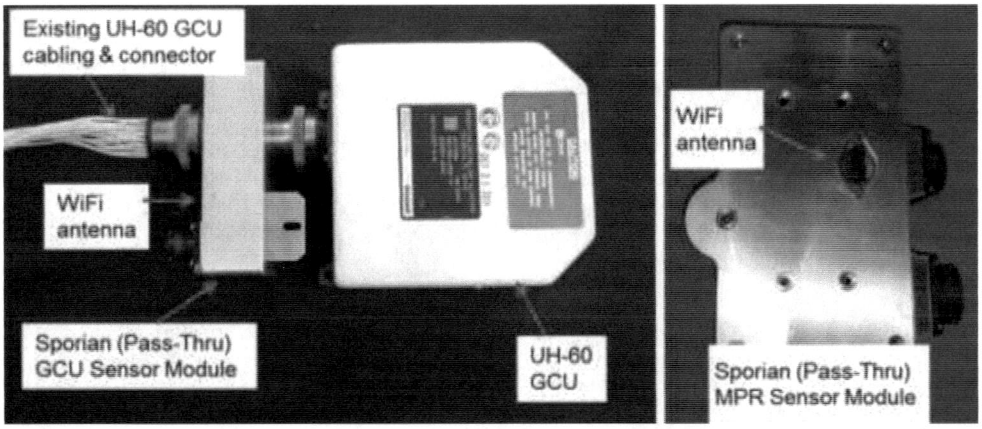

Figure 7: Fabricated GCU (left) and MPR (right) ESANodes in final H-60 form factor

A MIL-STD-810-oriented Vibration Test Suite has also been conducted. The tests included a resonance survey, sine dwells, a random noise dwell, and shock pulses. All levels, frequencies and durations were based on MIL-STD-810 F vibration for H60 Helicopters. An auxiliary wire harness and connector was creatd that allowed for the monitoring of all signals passing through the ESANodes during testing in order to continuously monitor for momentary or persistent opens during the tests. The data were collected at 25 ksps for the entirety of each test, four hour random noise tests included. The GCU and MPR Monitor devices were excited in three mutually orthogonal axes in compliance with MIL-STD-810 F.

Fit testing has been performed on UH-60L models at Colorado Air National Guard at Buckley Air Force Base (Figure 8). Sound proofing panels in the H-60 cabin area include a layer of aluminum sheet, so Wi-Fi transmission testing was also performed at Buckley. System on chip evaluation modules were programmed with special test code that could measure Wi-Fi performance criteria including Wi-Fi retransmission rate (Frame Check Sequence or FCS) and Wi-Fi signal strength. The maximum FCS error rate was around 13% which will cause some loss of transmission bandwidth due to retransmission of the bad frames. However, the modules were transmitting at a rate of 8Mb/s, which is four times higher than the target rate of 2Mb/s. The authors believe that the wireless communications concept for PowerIQ should be adequate for use in the H-60 rotorcraft with the current concept of operations.

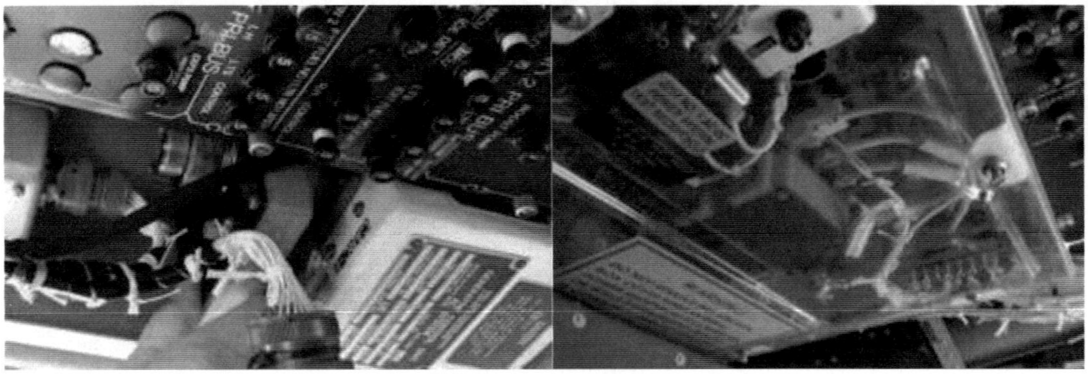

Figure 8: Fit checking 3-D printed GCU and MPR ESANode enclosures in an H-60L at Buckley AFB

Functional testing was performed with flight-certified GCU and IVHMU hardware in a representative, H-60 hardware in the loop (HIL) environment at UTAS.

The H-60 IVHMU includes the capability to communicate with a number of aircraft systems as required for maintenance and logistic support. A number of interfaces are provided to support these functions, including Ethernet,

MIL-STD-1553 buses, ARINC 429 interfaces, a USB bus, a CAN bus, and RS422/232/485 interfaces. HUMS data and data stored in the internal Cockpit Voice and Flight Data Recorder (CVFDR) may be accessed over Ethernet

The IVHMU is equipped with an Open Systems Interface (OSI) that enables integration of third party technology with minimal impact to IVHMU software configuration. OSI can provide a variety of services to internal and external clients such as the WBI. For the GCU/MPR monitoring application the IVHMU used an external Ethernet open system interface to communicate with the WBI. The WBI utilized three IVHMU OSI services: repository data, data logging and event logging.

Three primary test objectives included:

1) bi-directional interface with the IVHMU,
2) PowerIQ operation under nominal power system conditions (prognostic mode) and
3) PowerIQ operation under seeded fault conditions (diagnostic mode).

The hardware-in-the-loop test schematic is referenced as Figure 9. U.S. Army Research Development and Engineering Command, Aviation Development Directorate presided to observe the test.

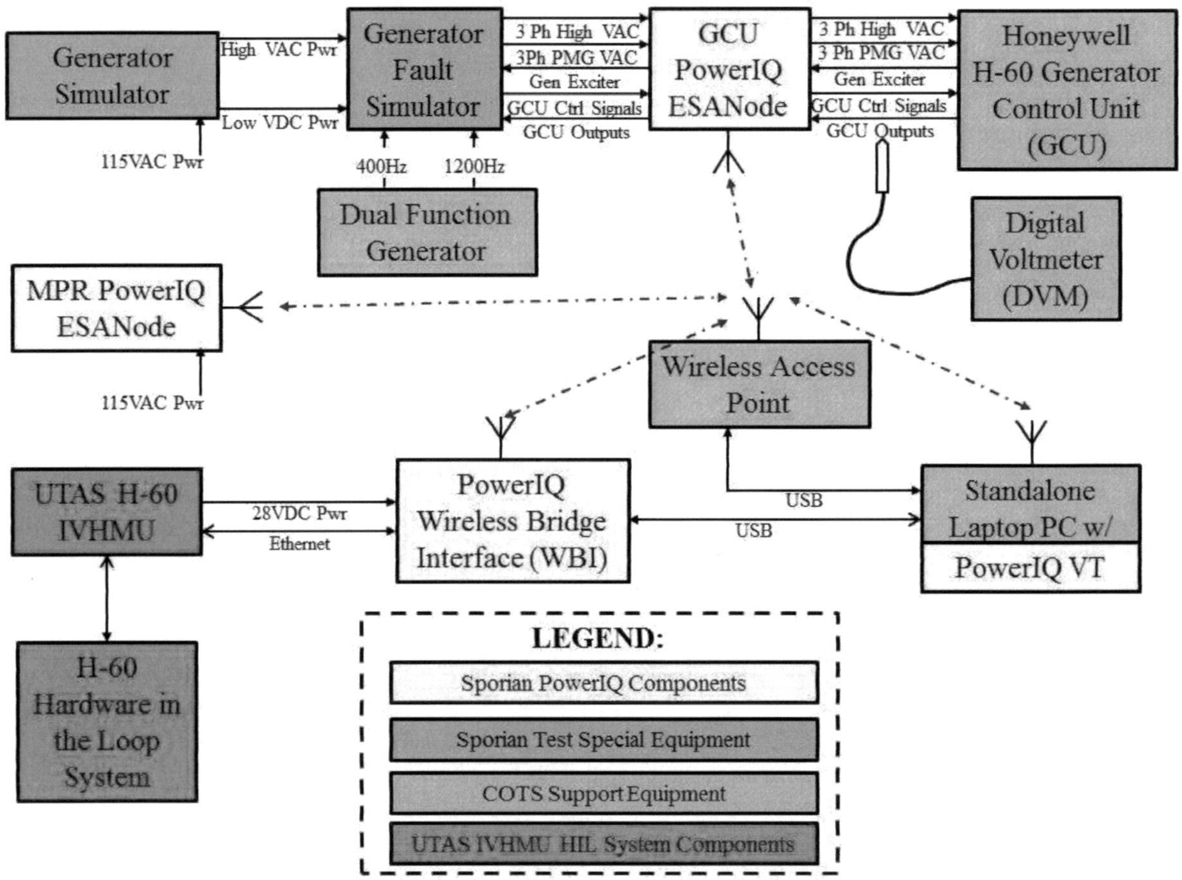

Figure 9: Schematic of bi-directional, seeded fault, hardware in the loop PowerIQ testing at UTAS

The IVHMU repository service was used by the WBI to request a pre-defined list of parameters from the IVHMU repository. The repository is the central working memory of the IVHMU and is continually updated as new data is received by the IVHMU. For the PowerIQ application, the WBI periodically requests system time and a set of parameters indicative of aircraft operation states. State information is essential to the PowerIQ concept of operations. For instance, the wireless concept of operations requires knowledge of the weight on wheels state.

The IVHMU data logging service is used to log prognostic signature data sent by the WBI. This data is stored by the IVHMU in a special data packet type called "private data". The private data packets with WBI parametric data are recorded to the IVHMU Memory Unit Download (MUD) file along with all the other IVHMU data packets. In an operational environment, the IVHMU ultimately transfers the MUD file to the H-60 Data Transfer Unit (DTU); this step was simulated in the HIL.

The event logging service is used by the WBI to log events in the MUD file. The IVHMU records these to the MUD file as an event packet which is readable by the IVHMU ground station. WBI generated events are identified as coming from the WBI.

After successful demonstration of bi-directional communications, the remaining two test objectives were pursued. Under normal operating conditions, power system data are captured in periodic snapshot mode and prognostic signatures are calculated. The system was operated in nominal mode for long enough to permit capture and logging of 3 data snapshots by the ESANodes. At that point, the HIL was used to simulate weight on wheels, the ESANodes subsequently transmitted raw data wirelessly to the WBI which logged the raw data and transferred prognostic signatures using the IVHMU logging services.

For the final, primary test objective, the system was operated under a total of three simulated electrical fault conditions: 1) high voltage over/under voltage, 2) permanent magnet generator under voltage, and 3) under frequency. Inducing these faults triggers the ESANodes to immediately capture a diagnostic event snapshot. Subsequent aggregation, signature analysis and IVHMU logging is similar to these steps under nominal operation. All three primary test objectives were successfully achieved.

A secondary test objective was to demonstrate various concepts of operation for non-HUMS equipped vehicles. One such concept is for an off-aircraft, handheld device such as an aircraft notebook (ACN) to communicate directly (wired or wirelessly) with the WBI to transfer raw diagnostic data providing an opportunity for further processing with VT. This concept was also successfully demonstrated. Signatures calculated from raw data with VT were compared with those calculated by the WBI and then transferred to the IVHMU as further validation of system operation.

Direct interface between WBI and a simulated H-60 DTU has also been demonstrated. The WBI has the potential to transfer files directly to the DTU on vehicles that are not equipped with HUMS.

Another approach for retrofit is to add removable mass storage to the WBI. Combined DTU and WBI functionality would then be implemented in the current DTU form factor.

A Windows software distribution CD has been developed that includes an executable copy of Visualization Tool. The distribution also includes captures from UTAS testing including nominal operations and seeded faults. Requests for the CD should be addressed to the coauthor from US Army Research Development and Engineering Command, Aviation Development Directorate.

A number of prognostic tools have been developed for the WBI. Data manipulation algorithms include root mean square (RMS) and peak-to-peak voltage and current calculations, power and power factor calculations, fast Fourier transform analysis, and Park transform calculation. State detection calculations include fundamental and harmonic frequencies, power spectral density, total harmonic distortion, and noise. These algorithms can also be implemented on off-aircraft systems including VT.

FUTURE WORK

Minor bug fixes and improvements are being incorporated into the GCU and MPR ESANode electronics and firmware. Systematic characterization over a wider array of fault scenarios is being undertaken and node-level stress screening tests are being conducted.

Unlike the ESANodes, the WBI is not currently in its final form factor (Figure 10). One significant system hardware and firmware development effort that remains is a redesign of the WBI into its final form factor. The next iteration of the WBI will be packaged specifically to retrofit on H-60M.

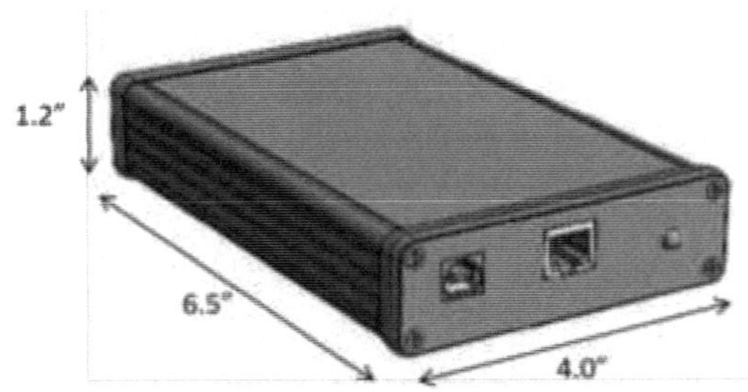

Figure 10: 1st Generation (COTS enclosure) WBI, CAD Model approximate dimensions (Left), Assembly (Right)

The authors are currently working with Sikorsky to down-select from potential H-60M mounting locations. Evaluation criteria include the availability of space and mounting provisions, proximity to needed IVHMU signals, and operating environment.

In the final version, the Ethernet interface between the WBI and the IVHMU will be replaced with a serial interface, though an Ethernet interface will be retained in the final design. Therefore, another major undertaking is to develop a RS485 interface with the IVHMU and demonstrate that interface at the UTAS HIL. This objective requires development of new code on both the WBI and the IVHMU.

Through HIL testing at UTAS, PowerIQ will have shown to successfully interface with flight-certified GCU and IVHUM hardware. Testing is also planned at the Sikorsky System Integration Lab (SIL) to demonstrate compatibility with flight-certified GCU, MPR and generator hardware with s simulated IVHMU.

This effort provides a unique opportunity to develop more robust prognosis for electrical systems components on H-60 and other rotorcraft. To aid in both system characterization and in the development of advanced prognostics, significant effort was expended developing a programmable, digital fault injector (Figure 11) to simulate a generator system under a wider variety of operating conditions. This second generation generator fault injector system (GFIS2) allows a user to simulate variations in voltage and frequency in both the high voltage generator and the Permanent Magnet Generator.

This capability will afford more realistic power system operation. For example, GFIS2 reads the Exciter output from the GCU and adjusts its analog outputs accordingly. In this way, it is possible to simulate a real-world load and measure how the GCU changes the Exciter output duty cycle in response. This helps demonstrate how the generator or the H-60 reacts to changes to the Exciter output.

GFIS2 was recently completed and functional testing is underway. This unit replaces the analog fault injector used for earlier testing at UTAS.

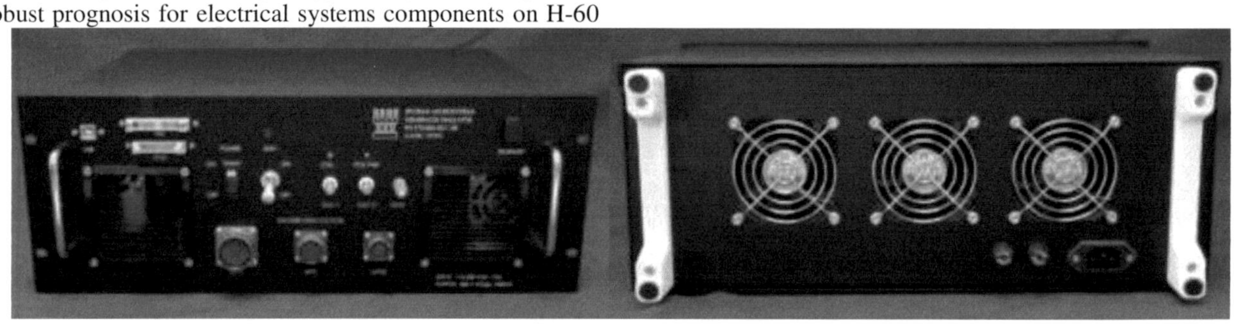

Figure 11: Generator Fault Injector Simulator, Front (Left), Back (Right)

Additional integration testing is sought at the US Navy Systems Air Command Propulsion Systems Evaluation Facility (PSEF) at Patuxent River.

Upon completion of functional, environmental and integration testing, Airworthiness Substantiation Documentation will be submitted to U.S. Army Research Development and Engineering Command, Aviation Development Directorate as a request for experimental flight testing.

CONCLUSIONS

A concept for a forward-fit and retrofit capable monitor for electrical system fault prognosis and diagnosis is presented. The system includes on- and off-aircraft components. On-aircraft components provide FIFO storage for raw diagnostic data and perform smart data reduction to support prognostic analysis. Off-aircraft tools developed to date include a signal visualization tool. A realization of this system for H-60 is under development using the spiral method. Significant risk reduction for the H-60 realization has been accomplished through functional testing, environmental stress screening, and systems integration demonstrations. At the conclusion of the planned effort, a prototype will be a strong candidate for experimental flight testing and further efforts to transition to fielded rotorcraft.

ACKNOWLEDGMENTS

This material is based upon work supported by the U.S. Army Research Development and Engineering Command, Aviation Development Directorate under Contract Numbers W911W6-13-C-0034, W911W6-14-C-0019 and W911W6-16-C-0023. Any opinions, findings and conclusions or recommendations expressed in this material are those of the author(s) and do not necessarily reflect the views of the US Army.

The authors also wish to thank Sikorsky Aircraft Corporation, United Technologies Aerospace Systems, Colorado Air National Guard and Wisconsin Air National Guard.

REFERENCES

[1] Wilmeth, Randall G., and Michael W. Usrey. "Reliability-centered maintenance: A case study." *Engineering Management Journal* 12.4 (2000): 25-31.

[2] Ginart, Antonio, et al. "An L 1 norm for efficient power quality monitoring." *Electric Power Systems Research* 79.11 (2009): 1495-1502.'

[3] Kalgren, Patrick W, Carl S. Byington, et al. "Automated Data Translation for Avionics Health Management through Intelligent Evidence Streams and Databases." *AHS Annual Forum*. 2004.

[4] Kalgren, Patrick W., Carl S. Byington, and Michael J. Roemer. "Defining PHM, a lexical evolution of maintenance and logistics." *Autotestcon, 2006 IEEE*. IEEE, 2006.

[5] Stewart, R. M. *Some useful data analysis techniques for gearbox diagnostics*. University of Southampton, 1977.

[6] Lebold, Mitchell, et al. "Review of vibration analysis methods for gearbox diagnostics and prognostics." *Proceedings of the 54th meeting of the society for machinery failure prevention technology*. Vol. 634. 2000.

[7] Ginart, Antonio E., et al. "On-line ringing characterization as a PHM technique for power drives and electrical machinery." *Autotestcon, 2007 IEEE*. IEEE, 2007.

U.S. Army AED Aviation Data Science 2018 Update

Dr. Andrew Wilson
Daniel Wade
United States Army
Aviation Data Science Team
Redstone Arsenal, AL, USA

ABSTRACT

Over the past several years, a focus on and vision for data science in operational aviation data sets has emerged from the United States Armys Aviation Engineering Directorate (AED) team that has traditionally focused on vibration diagnostics and HUMS applications. Recently formed into a separate team under the Aerodynamics and Simulation Branch, this team has pursued the organizational, technological and intellectual challenges required to apply data science to aviation data, and to field data science products to systems under airworthiness governance. This paper provides an update on several of the specific areas of accomplishment and direction, with further detailed information referenced.

AVIATION DATA SCIENCE TEAM

The US Army is the worlds largest operator of rotorcraft. This fleet generates a very large amount of operational data. Some portionsthe maintenance logs, the overhaul records, the flight logsare analyzed by various organizations for purposes such as business intelligence and reliability inference. However, by volume (i.e. gigabytes), the largest portion of the data is the Health and Usage Monitoring System (HUMS) data generated on board the aircraft themselves. This dataset comprises some 20+ terabytes of what is, in essence, flight test data for most operational flights over a period of 10+ years.

The dataset has already been used for some very significant engineering and business improvements to the platforms. Significant cost and safety gains have been made by adjusting part lives using information about how the aircraft are actually flown. Enormous progress has been made in understanding how vibration diagnostics truly reflect mechanical health condition.

However, such applications have been piecemeal and cumbersome. It can take weeks to months to organize, query and pull data sets suitable for either traditional engineering analysis or machine learning techniques, followed by an equally large period of time to perform the analysis. Nearly all analyses require querying multiple data sets, such as the maintenance logs and the HUMS data for part lifing, and the HUMS data with the depot overhaul teardown records for vibration diagnostic improvements.

Further, there are multiple reasons that modern data science methods require adaptation and caution when being applied to aviation and other critical-safety domains. Charged with governing the airworthiness of Army aircraft, the AED is responsible for ensuring both that the aircraft continue to be safe, and also with ensuring that the Army is able to field new technologies safely. It has become clear that AED must provide qualification guidance as new data science-driven systems gain momentum in the industrial base which serves Army Aviation.

The AED Data Science team was created this past year to address these data related issues, which are certain to grow at an ever-increasing pace. Organized under the Aerodynamics and Simulation branch of the Aeromechanics Division, the Data Science team has two core competency areas.

One area is data science for legacy data. It is important to ensure that the vast information that has been collected on the existing legacy platforms is used to maximum extent to inform engineering decisions, both for improvements to those legacy platforms and to inform requirements for future systems. This is accomplished by being subject matter experts (SMEs) for at least one type of aviation data, and by forming coalitions with other data SMEs to determine what significance each kind of data has for the others.

Another core competency is establishing verification and validation (V&V) requirements for data science models to be used in support of airworthiness qualification. Data science has its own unique set of powerful possibilities and dangerous misuses. V&V requirements look different for data science models than for physics-driven mathematical or computational models.

To further these core roles, the Data Science team has taken several steps. First, we have pursued the training and practice of data science. We have authored papers along the way describing these efforts (References 1-8), with some recent updates provided here. We have also established ongoing collaborations with different government and industry groups. Several collaboration projects have concluded, others are ongoing, and some are just beginning.

Presented at the AHS Airworthiness and HUMS Specialists Meeting, Huntsville AL, February 23-25 2018.

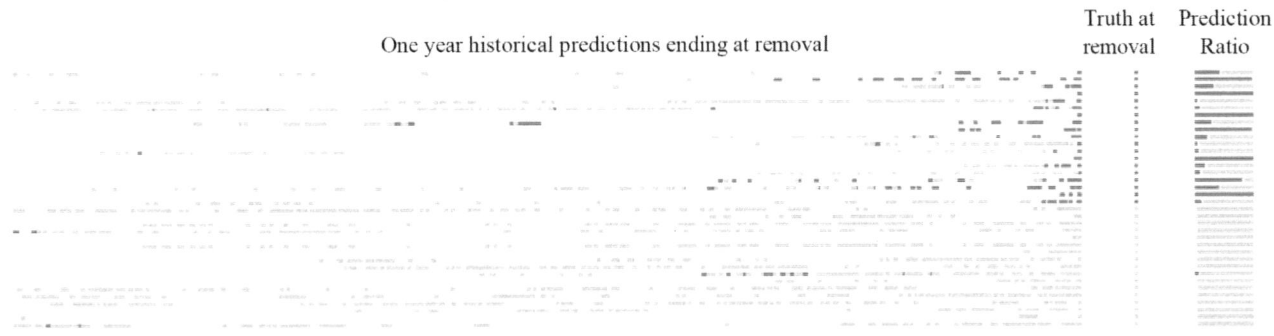

Fig. 1. Example visualization of Historical and Asset Based TNR ratio calculation for a logistic regression model learned from the NGB training vault data.

MACHINE LEARNING ON A NOSE GEARBOX

For the past three years, the Data Science team has been pursuing a flagship machine learning project, to simultaneously demonstrate the application of machine learning to mechanical diagnostics, and to develop a process for certifying machine learning diagnostics for airworthiness applications. The project was to develop a machine learned diagnostic classifier which provides the maintainer with a single healthy/faulted indicator to detect nose gearbox faults. Efforts along this path have been previously described (Ref. 1). Since those updates, several things have happened, including bringing the project to a successful conclusion.

First, Honeywell Aerospace joined the project, developing several machine learned classifiers to compete against those algorithms the Data Science team had already employed. These models compared very favorably to the best models produced at the Army.

Second, the best models from both the Army and Honeywell were selected and permitted to gain access to the formal validation set (called the "pretest set" in previous AED data science papers; "validation" is the more standard data science term which the Aviation Data Science team has now adopted). Detailed metrics from this validation set were compared with the in-sample and cross validation metrics computed from the training set. Figure 1 shows an example of how these models were evaluated, using not just the measurements near removal but for the history of gearbox installation. From these models, a single model was selected as the winning model, which was in turn evaluated over the final test set. This process was previously described in detail (Ref. 2).

This final, winning model was one of those developed at Honeywell. It was determined that the Honeywell winning model met or exceeded the performance of the existing, heuristic health model (which takes the worst threshold exceedance among a number of condition indicators). This model was deployed at the enterprise level to monitor gearboxes on an ongoing basis, and provided regular exceedance reports.

While this model did succeed in marginally improving performance, and clearly succeeded in reducing a set of indicators to a single indicator, the project was much more useful for the lessons learned.

Among these lessons: data sets are somewhat living things. Do not expect that the first version cleaned to a format suitable for machine learning will be the final version. As model builders work with the data, more and more issues can come to light. Having multiple people with very different perspectives working on the data, in detail, can help to illuminate these issues, which include how to scale each variable, and how to curate "faultedness" to be consistent and learnable.

The Army also stumbled across an important lesson in the process of developing the crawler (Ref. 1), a technique the Aviation Data Science team used to search for good nose gearbox health models. To choose the best possible models and tuning parameters for those models, a computer program was developed that randomly selected: a model, each tuning parameter for that model, and a subset of input columns to use for training. Then, it performed a cross-validation on the training data, and finally trained the algorithm on all the training data, saving the final model and a set of performance results (accuracy, informedness, true positive rate, etc.). The resulting collection of models was then be searched for those which performed the best.

This form of random search is a fully valid, albeit potentially computationally expensive, optimization technique. However, besides the random choices intended, the crawler also randomly assigned examples to different cross validation folds differently for every training run. The result: the highest performing models all had very similar example assignments in folds. In essence, as one analyst put it, we had taught the machine how to cherry-pick the data to make the training and cross-validation look optimal. The fix was to perform a single, random assignment which allocated each training example to a cross-validation fold permanently, and report cross-validation performance across models over this fixed distribution of examples. The broader lesson here is: when performing optimization, especially very complex optimization, it is important to double check that what your code is optimizing what you think it is optimizing.

Finally, and most importantly, the work on the nose gearbox project led to the creation and revision of a process whereby machine learned classifiers can be placed into service in support of airworthiness, as published previously (Ref. 2).

DIRT BIKE SLOPE SPECTRA

Briefly, a new form of data quality failure was observed for a particular, newer accelerometer installation. The spectra looked somewhat like the classic DC offset spectra, with a large scale decline, except that the low frequency portion rose from zero to a peak instead of falling off like a ski-slope. This new shape was dubbed "dirtbike slope", and was noticed by an analyst looking at data manually (see Figure 2).

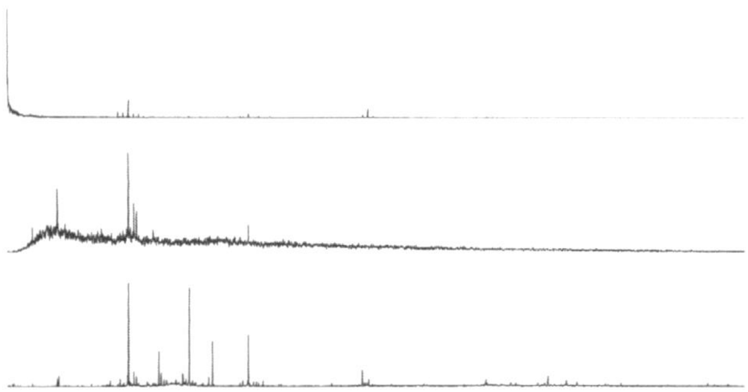

Fig. 2. Example spectra: ski-slope fault (top), dirtbike slope fault (middle), nominal/typical (bottom).

In order to search the fleet for similar spectra, a simple machine learning classifier was trained on the few positively identified cases, along with manually labeled "nominal" cases. The most strongly identified positive cases were submitted to the analysts for review, who classified them as nominal, ski-slope, dirtbike slope, or other. Then a new classifier was trained on the larger data set.

This whole process was repeated 5 or 6 times, until the results were strongly consistent and several new cases had been identified, including a case which had not yet been corrected (accidentally or intentionally). This enabled the prime analyst to pursue possible causes on the aircraft which was experiencing the problem (Ref. 1).

ZOMBIES AND THE IMPORTANCE OF METRICS

In the course of developing the Nose Gearbox diagnostic classifier, the problem of severely skewed datasets came to light. Classification datasets which are skewed have classes (such as healthy) for which there are many more examples than other classes (such as faulted). Mechanical health data for aviation is severely skewed, having healthy to faulted measurement ratios in excess of 10,000.

This issues became a major challenge, because with such severe skew ratios classifiers can be 99% accurate by always guessing healthy. Accuracy is, thus, a bad metric. There are a number of metrics which attempt to deal with this problem, such as sensitivity, specificity, F1 score, and others. As a single roll-up metric which expresses the overall value of a classifier, each was found inadequate for various reasons. After discovering a paper by Powers (Ref. 9) which includes a detailed development of the impact of skew (called prevalence) and model preference (called bias) on many different metrics, the metric of informedness was selected as a high-quality metric for rating classifiers over datasets of different skew ratios.

Informedness is simply the true-positive rate minus the false-positive rate. This metric, while oddly simple, has several properties which make it powerful and useful. Regardless of the skewness of the underlying populations, informedness can be interpreted as the probability of beating the true baseline model: random chance. A model which flips a possibly weighted coin will, on average, have an informedness of 0; a model which has an informedness of 1 has perfect accuracy; and two models of informedness 0.5 will both beat random chance 50% of the time, though they may make the true-positive/false-positive trade-off in different ways. All of these properties hold regardless of the population skew.

It became difficult, however, to communicate to engineering and even data science practitioners why informedness was a good metric, and why other metrics were lacking. The Powers paper develops the math in great detail and accuracy, and perhaps because of this is not the easiest to digest.

Further, there is a lack of perceived *need* for better metrics with more clearly understood properties. For many machine learning applications outside of critical safety, the question is mostly academic. If a stock prediction algorithm is 51% right, for some reasonably cheap metric like weighted accuracy, then it will make money. If an image classification algorithm is wrong 10% of the time, then a user received a slightly inaccurate targeted ad. But for aviation and other critical safety applications, the difference between different kinds of errors *matters*, for the eternal balance of risk and cost.

Hence, the need to develop clear guidance on how metrics impact decision making concerning classifier models. To do this, the Data Science team worked on a toy problem with a time-dependent skew: the zombie apocalypse. Early in the zombie apocalypse, there are very few zombies. But as the infection spreads, the skew changes, until zombies and people are in balance (numerically, at least). As humanity dwindles, the skew ratio changes still further. So what is the consequence in these different circumstances of using different metrics to make decisions about a zombie/human classifier algorithm?

This problem, while meant to be humorous, has many of the properties of the problems critical safety practitioners face. The consequences of type I and II errors are significant to the extent of human life and liberty.

The results of this work are nearly complete, and will be presented at the ASME Verification and Validation Symposium in May of 2018.

A MAIN TRANSMISSION DIAGNOSTIC

For various reasons, one of the main rotorcraft transmissions the Army monitors with on-board vibration diagnostics does not have any validated condition indicators. Some of these reasons remain valid, such as suboptimal sensor placement and incomplete historical data. However, the spectrum reconstruction work (Ref. 3) changed the potential value of many of the historical tear-downs. Upon curating the available tear-down and sensor data, this particular epicyclic gearbox now has a significant number of documented faults paired with operational vibration sensor data. With the development of a machine learning process to build a diagnostic, it became a question: can we assemble a good dataset and train a diagnostic?

In the case of the Nose Gearbox diagnostic described above, the Army already had validated condition indicators which correlated strongly with available ground-truth evidence. For this main transmission, there is no validated diagnostic. Any model with reasonable discriminatory power would be a significant improvement.

The available ground truth records were compiled and correlated against the available HUMS data. Training, Validation and Test sets were established, and gearbox installations and their associated data were assigned to each of them. Several model practitioners then used the training data to develop models, reporting in-sample and cross-validation performance. In short, the process developed during the Nose Gearbox project was applied to a Main Transmission.

This project has not yet concluded, but has some promising initial results. Pending review of the validation results, a diagnostic may be tested and fielded in the coming months.

DEEP LEARNING A SURROGATE MODEL FOR VIBRATION

Surrogate modeling seeks to model not some unknown physical process, but some known mathematical or computational model, but with enormously reduced computational overhead. Over the past year, the Data Science team has explored several applications of deep learning for surrogate modeling of various problems in aviation.

Jointly with a group at Sandia National Laboratory, deep learning was applied to two different dynamics problems with similar model structures. On the Sandia side, the technique was applied to surrogate model the behavior of near-wall turbulent boundary layers. In both cases, the inputs and outputs were spectra. The Army model predicts one vibration spectrum using the spectrum captured from a perpendicular sensor, as trained using field data. The Sandia model predicts the wall pressure power spectral density (PSD) using a distant PSD, as trained using high-resolution direct numerical simulation (DNS).

Figure 3 shows an example of the results achieved by the Army model. The spectrum in red is from a sensor oriented horizontally, and the spectrum in blue is from a vertically mounted sensor, mounted on the same sensor bracket. The model is designed to take the red spectrum, and attempt to predict the blue spectrum, with its results shown in green. This is a randomly selected spectrum from a validation set, i.e., this spectrum was not used by the model during training. The accuracy is substantial, and demonstrates how this model could be used in place of the actual sensor data. Future uses for this kind of model include eliminating unnecessary redundant sensors on board aircraft, or detecting sensor or mechanical anomalies by comparing predicted to actual spectra.

This work was substantially successful, showing the potential of surrogate models for reducing computational burden in coupled fluid-structure solvers (on the Sandia side) and the potential of reducing the number of sensors on an aircraft (on the Army side). This work was presented at ASME's Verification and Validation Symposium (Ref. 4).

AMRDEC/ACE DATA SCIENCE COLLABORATION

The Data Science team has joined with elements from multiple directorates within the Aviation and Missile Research, Development and Engineering Center (AMRDEC) and elements of the US Army Corps of Engineers (ACE) Engineer Research and Development Center (ERDC) to form an aviation data science collaboration.

The objective of this collaboration is to push forward the DoD HPC tools for data science, while applying data science to concrete problems faced by Army aviation. To date, the major accomplishments include moving the entire history of operational data to a DoD supercomputer, adapting existing tools to enable embarrassingly parallel processing of that data, and adaptation of existing data science projects to run on DoD supercomputers.

This is an ongoing effort that will spawn multiple updates over the coming months.

FINAL THOUGHTS

This paper is intended to provide a series of smaller updates, so the community has a clearer picture of the efforts going on within the AED Data Science team. The Data Science team grew out of HUMS and vibration diagnostics, and continues to pursue improving those systems and techniques. However, the scope of application of data science techniques is continually broadening, with applications to structural part lives, flight controls anomaly detection, and natural language processing coming in months, not years. As the uses of the legacy data expand, and the ways in which the legacy fleet data informs future platforms improves, the Data Science team will continue to evolve to make the best use of the incredibly valuable collection of data compiled by the US Army on rotorcraft operations.

REFERENCES

[1]Wilson, A., Wade, D., Albarado, K., and Partain, M., J. andStatham, "A Classifier Development Process for Mechanical

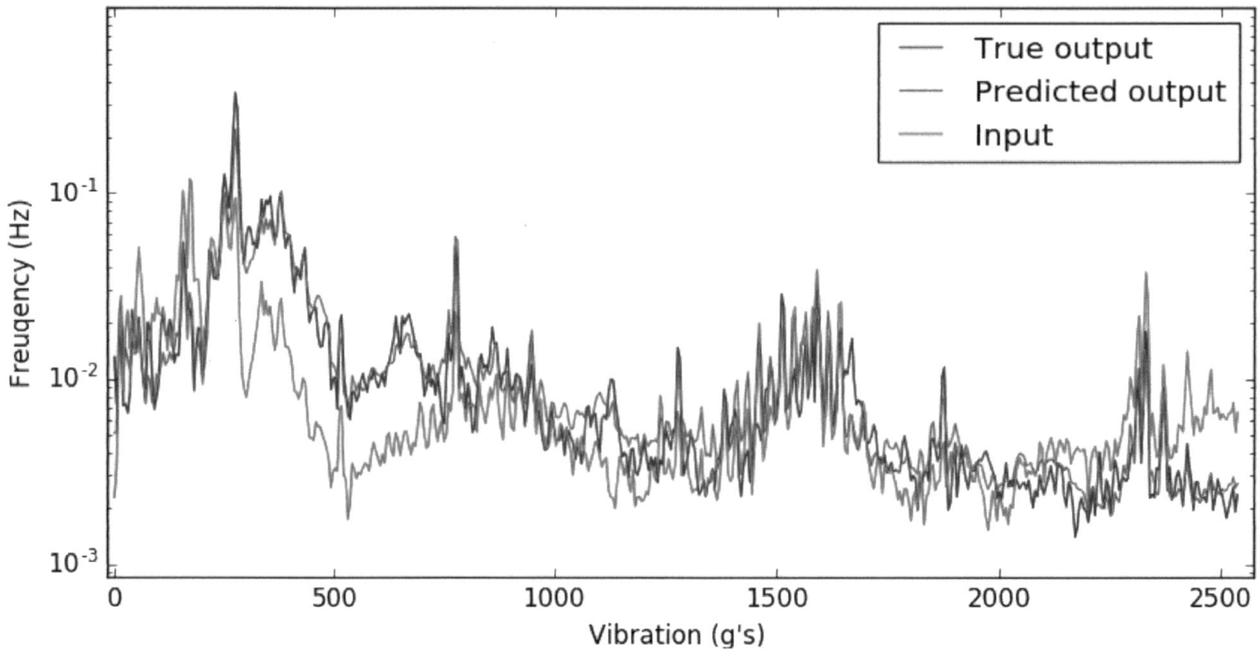

Fig. 3. Example output of Army perpendicular sensor model, compared with the true value for the spectrum and the input spectrum.

Health Diagnostics on US Army Rotorcraft," *ML and PHM Workshop, SIGKDD 2016*, 2016.

[2]Wade, D. and Wilson, A., "Applying Machine Learning-Based Diagnostic Functions to Rotorcraft Safety," *Tenth DST Group International Conference on Health and Usage Monitoring Systems, 17th Australian Aerospace Congress*, 2017.

[3]Wilson, A. and Wade, D., "Reconstructing Spectra from IVHMS Condition Indicators," *Proceedings of the 73rd Annual Forum of the American Helicopter Society*, May 2017.

[4]Wilson, A., Wade, D., Ling, J., Chowdhary, K., Davis, W., Barone, M., , and Fike, J., "Convolutional Neural Networks for Frequency Response Predictions," *ASME Verification and Validation Symposium*, May 2017.

[5]Wade, D. and et. al., "Measurement of Vibration Transfer Functions to Inform Machine Learning Based HUMS Diagnostics," *Proceedings of the 72nd Annual Forum of the American Helicopter Society*, May 2016.

[6]Albarado, K. M., Wilson, A., and Wade, D., "Advanced Neural-Networks for Mechanical Health Diagnostics," *Proceedings of the 5th American Helicopter Society CBM Specialists Meeting*, February 2015.

[7]Wade, D., Vongpaseuth, T., Lugos, R., Ayscue, J., Wilson, A., Antolick, J., Brower, N., Krick, S., and Szelistowski, K., M. andand Albarado, "Machine Learning Algorithms for HUMS Improvement on Rotorcraft Components," *Proceedings of the 71st Annual Forum of the American Helicopter Society*, May 2015.

[8]Wade, D., Lugos, R., and Szelistowski, M., "Using Machine Learning Algorithms to Improve HUMS Performance," *Proceedings of the 5th American Helicopter Society CBM Specialists Meeting*, 2015.

[9]Powers, D. M., "Evaluation: from precision, recall and F-measure to ROC, informedness, markedness and correlation," *Journal of Machine Learning Technologies*, Vol. 2, 2011, pp. 37–63.

Dynamic Time Warping for Pattern Detection in Time Series Flight Data

Matthew Statham
Dr. Andrew Wilson
Daniel Wade
United States Army
Aviation Engineering Directorate
Aviation Data Science Team
Redstone Arsenal, AL

ABSTRACT

Rotorcraft with HUMS systems typically generate a large amount of time series data such as ambient air temperatures or various engine temperatures. For a fleet over time, this data set can become quite large and can't all be consumed by the human eye in the form of strip charts in a timely manner. When an anomaly occurs, there is a likely chance that a subject matter expert (SME) will want to find out if that particular anomaly has ever happened before and when it happened. The SME will want to find anything in the history of the fleet that looks like the anomaly that they just saw. This paper presents a method called dynamic time warping (DTW) for doing just that. A SME can input a query array and the DTW algorithm will find all occurrences that are similar to the one the SME input. The result is a drastically reduced set of data for the SME to physically look at. DTW will be introduced and explained and an example use of DTW on time series flight data will be presented along with tips and tricks for expanded use of this type of algorithm.

INTRODUCTION[1]

A seemingly common task when dealing with time series data is finding occurrences of an identified anomaly in an automated fashion. In the area of aviation data science, the anomalies can be a particular event such as a hard landing or an engine over-speed. The time series would be the parametric data taken from the aircraft such as temperature, altitude, or rotor speed. Searching through thousands of flights, though, can be rather problematic. The goal of this paper is to introduce a method that has typically been used in the realm of computer science to the world of aerospace engineering by applying dynamic time warping (DTW) to time series flight data. The Army's Aviation Engineering Directorate, Data Science Team, is developing algorithms that automatically identify events in time series data to improve Health and Usage Monitoring System capability. Developing these capabilities can increase engineering efficiency, reduce time required to perform requested analyses, and potentially increase the accuracy of analysis.

THE CLASSIC APPROACH: THRESHOLDING

The typically approach when asked to find a known occurrence in time series data is various types of thresholding. One would look at the occurrence they want to find and try to come up with maximum and minimum values that may describe this occurrence in other scenarios. While this can be effective sometimes, it's far from perfect when considering flight data. This method is not robust against an anomaly that say, for example, has a slightly higher magnitude or maybe this time the anomaly happened to last for slightly longer than usual. In addition, writing this code is highly specialized to a specific problem and can take a while to develop.

COMPARING ARRAYS: DISTANCE AS A SIMILARITY METRIC

When searching through data using an example, the essential question that must be addressed is: how *similar* is the *query* to the *data.* In the case of time series data, this is asking how *similar* or *dissimilar* two different time series are to each other.

Assuming both series are the same length, a standard way of doing this is using a root-sum-square (RSS) difference of the two time series as vectors. This can be thought of as an RSS error, but it is also often called the Euclidean distance between the two vectors. This RSS distance or error is a *dissimilarity* measure: the larger it is, the more dissimilar two vectors are. When it is zero, the two vectors are identical.

But this method can only give accurate results only in a very controlled scenario. Controlled and predictable are two things that real flight data are not. This method is not robust against an anomaly that say, for example, has a slightly higher magnitude or maybe this time the anomaly happened to last for slightly longer than usual. These difference would increase the distance metric significantly, whereas a human

---

[1] Presented at the AHS Airworthiness and HUMS Specialists Meeting, Huntsville, Alabama, USA, February 21-22, 2018.

eye would be able to easily identify the anomaly that was being searched for.

DYNAMIC TIME WARPING

Dynamic Time Warping (DTW) is a method that measures *likeness* of two time series arrays. DTW, originally used for speech pattern recognition (Ref. 2), calculates a distance metric that is a more appropriate method by which to measure likeness because it maps like features. Calculating distances between like features makes this method more robust against changes in time shifts and minor shape changes.

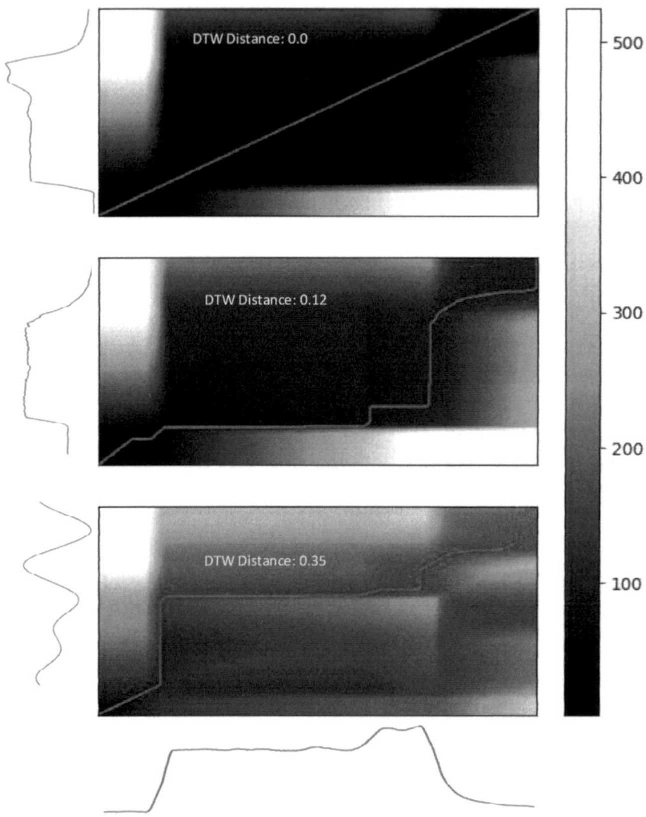

Figure 1: DTW Calculation Examples

Figure 1 shows three examples of DTW calculations. Each of the three plots on the vertical axis is being compared to the plotted array at the bottom of the figure on the x-axis. The top plot shows what DTW looks like if the same two arrays are compared. The most efficient mapping path is a diagonal line with no difference between the arrays along this path. The calculation of the two arrays creates a two-dimensional matrix where each element represents the distance between each point of one array to that of the other array. As such, the blue line consistently goes through dark regions as these regions have the smallest comparative distance. Conversely, the mapping of the bottom right and the top right of the plot shows the distance mapping between two different sides of the arrays which are basically useless.

The second plot shows an array that is somewhat similar but would not be considered a match. Note how the line goes off the 45-degree line but doesn't get very light. The dark color implies that the distances encountered are not very high.

The third plot is a comparison of a combined sine wave generated for comparison purposes. Note that the plot goes significantly off of the diagonal and the overall color is much lighter, indicating a much greater dissimilarity (or error, or distance) between the two time series.

STREAM MONITORING

Simply comparing two equally sized arrays is useful, in some scenarios, but not when searching for a particular transient in a collection of long time series. The goal here is to find a particular section in a time series that looks like the known *event* that is being searched for. This known event is defined as the **query**. The array being searched is defined as the **target**. An assumption that is made is that the query is much shorter than the target. The approach is to use a sliding window over the target array. More specifically, evaluate the time-warped distance between the query and a *query-sized slice* of the target array at each time step in the target array. Figure 2, below, illustrates this process.

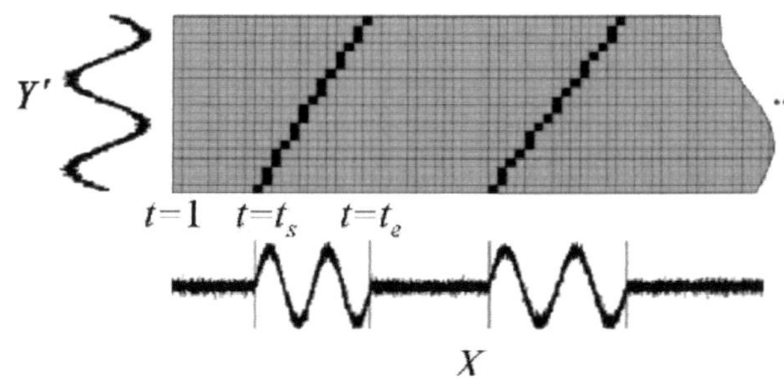

Figure 2: Stream Monitoring with DTW. Taken from [1]. Y' is the query, X is the target, ts is the start time of a DTW comparison, and te is the end time of a DTW comparison.

As the process takes place, DTW distances for the current and previous time steps in the target query are saved. Alongside these distances, is the time step (array index) in which the minimum path ends at that point. This is important because if the distance reported at the top of the array is less than the threshold, the origin of the result has to be reported. This specific algorithm is call SPRING and can be read about in more detail in Reference 1. Other similar techniques have been used for stream monitoring as well.

EXAMPLE: ENGINE OVERSPEEDS

The example that will be presented is automated identification of engine overspeeds on a rotorcraft turboshaft engine. This type of overspeed is referred to as a control transition overspeed. What is physically happening in this type of overspeed is not important. What is important is that an SME saw something in the data and wanted to know if it had ever happened before and when. For completeness, the

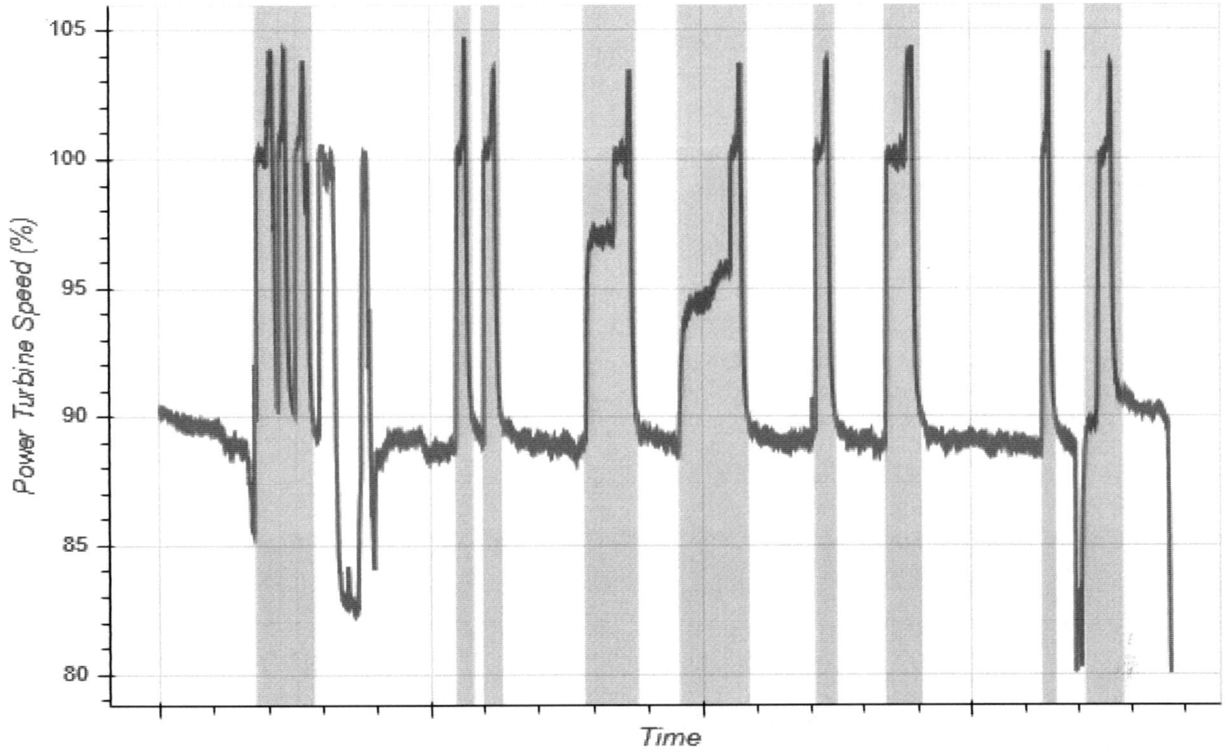

Figure 3: Engine Over-speeds Identified by DTW

parameter that was searched was power turbine speed (NP). Visually identifying flight events such as engine over-speeds in flight data can be rather straightforward because of their shape and magnitude, but writing an algorithm to identify them can be rather difficult.

One down side to this overspeed example was that there was only one concrete flight example to use as the query source. Figure 3 below shows the one flight that had multiple over-speeds in it, which are highlighted in red by the DTW algorithm. Note that all of the examples share the same shape but often are wider, taller, or have a smaller first "hump". The query was taken from this array and used to identify the others. The two peaks that do not have this shape were not flagged as overspeed conditions.

Almost all papers on DTW will say to Z-Normalize the data for both the query and the targets so that the mean value is zero. This makes sense because if everything has a 0 average, the algorithm will focus more on shapes rather than magnitudes. However, if the variable that is being monitored is one such as rotor speed or engine power turbine speed, Z-Normalization may not be necessary and may hinder performance. Variables such as these are usually very consistent between aircraft. Rotor speed, for example, is almost always going to be around 100% for a conventional rotorcraft

NOTES ON VIBRATION DATA

The original investigations into DTW was related to finding trends in HUMS vibration condition indicators (CIs) over time. The conclusion from this investigation as that the data was in general too noisy even after a great deal of filtering. In regard to vibration spectra, this technique could potentially be used to compare sections of downsampled spectra.

COMPUTATIONAL PERFORMANCE ISSUES

The naïve DTW comparison algorithm is $O(n^2)$ in space and time where n is the number of points in two equally sized arrays that are being compared. This makes comparison of even reasonably small time series impractical. Many methods have been developed to ease the time and space pitfalls such as narrowing the window in which the DTW mapping can stray so the entire n^2 matrix doesn't have to be calculated and stored.

The SPRING algorithm eases the time and space computation burden significantly. Only two arrays of the size of the query have to be stored and the time is O(nm) where m is the length of the query. This can still be rather cumbersome especially for longer flights or if many flights have to be scanned. While the actual speed is very hardware dependent, an example performance seen for single variable screening was about 63,400 flights in 48 hours where both engines on each flight were scanned. This was multiprocessing with four processes in python on an Intel Core i7 processor. This is still cumbersome considering that this represented only six months of data for a fraction of the Army's fleet. Also, consider that not all of these flights had to be scanned because of a few early abandoning techniques involving simple max thresholding filters.

To ease these computational burdens, a few techniques can be used. First, the process of searching multiple flights is embarrassingly parallel or in other words, one flight's search doesn't depend on another flight's search. For this reason, multiprocessing is trivial. Second, various prescreening techniques can be used so that the DTW distance doesn't even have to be calculated. For example, our propulsion SMEs told us that they wouldn't consider an overspeed that was identified by DTW if it didn't exceed 104% NP. These arrays were discarded before being screened. Finally, downsampling is essential. For the example above, decimation was used after an 8th order Chebyshev filter was applied to the signal. The decimation factor was 4. The same process was applied to both the query and the target. The biggest consideration is that the query has to maintain its shape and be on the same order of decimation as the target so the two are still comparable.

FUTURE WORK

In regards to the efforts in Army Aviation, DTW stream monitoring has only been used to try and detect engine over-speeds on a fraction of the Army's data in only a single variable sense at the time this paper was written. There are many other anomalies that could be scanned for such as hard landings or maybe auto-rotations if that is of interest. A large advantage of this technique is that it's relatively general and can be applied to many different scenarios with really only having to modify hyper parameters and not the actual algorithm. Historically, at least for the Army, time series parameter data hasn't been utilized very effectively. This technique could help in making this data more useful.

As mentioned earlier, there are likely improvements that could be made regarding computational affordability which would make this technique quicker to deploy. There are techniques used for single variable naïve DTW comparisons such as banding the calculation window and some early abandoning techniques that could potentially be used in stream monitoring for example. If this is done, then more multivariate processes could be more effectively explored.

DTW can also be used in different ways. DTW can be very effective at finding averages between multiple arrays using a technique called barycenter averaging. As an extension to the averaging of arrays, this could be used to do clustering on different types of arrays in much the same way that clustering algorithms could be used for scalar values. This would allow a user to "group" different types of time series arrays.

REFERENCES

[1] Y. Sakurai, C. Faloutsos, M. Yamamuro, "Stream Monitoring under the Time Warping Distance". National Science Foundation

[2] A. Mueen, E. Keogh. "Extracting Optimal Performance from Dynamic Time Warping". 22nd ACM SIGKDD

Algorithm Suite for Rotorcraft Bearing Fault Detection

David Noever
Senior Technical Fellow
PeopleTec, Inc.
Huntsville, AL US

ABSTRACT

To identify and classify mechanical faults, the modern tools of data science transform raw vibration data into key features from which either thresholding or learning algorithms can make predictions for damaged components. Their success relies on a complex chain of pre-processing spectral inputs. The aviation community has placed less emphasis on finding the best algorithm. We systematically evaluate 35 classifiers representing 9 major algorithmic families against features extracted from the University of Cincinnati bearing dataset. We compare the classifiers based on statistically significant differences in accuracy and execution time. Ensembles of boosted decision trees provide the best classifier family for diagnosing faulted bearings with 98% accuracy. In this study, currently favored methods which threshold single features for condition indicators offer an example of a simple one-branch partition tree, which performs significantly worse (77%) than most other classifier families.

INTRODUCTION

Current machine learning enthusiasm stems partly from vastly larger datasets and more computing power (including graphical processing units), but has similarly spawned rapid growth in algorithm development. Large algorithm libraries (in Matlab, R and python particularly) now suggest new research in: 1) best classifier discovery (even without extensive domain knowledge or parameter tuning); 2) common scoring metrics to compare the different methods fairly; and 3) large algorithm ensembles which may perform below-average on some features but combine to a state-of-the-art decision maker. This work follows similar comprehensive approaches in Refs. 1-2 for comparing many learning algorithms (179) across diverse datasets (121), but where possible specializes the discussion for bearing faults to identify particular weak and strong performers for both algorithm families and their key predictive features.

Condition based maintenance aims to predict faults and thus minimize costly, unscheduled repairs. Sensors placed on key components collect vibration data and the largest contributing spectral features offer a path to identifying troublesome behaviors. From an algorithmic perspective, setting a caution threshold acts as a single-branch decision tree. For example, if a high spectral energy is associated with a resonance of another moving part, the signal may indicate an imminent gear, bearing or shaft failure. The net outcome may capture a high percentage of known failures. But to identify the remaining outliers, other data science fields (see Ref. 2) have found recent successes with combining multiple algorithms (e.g. voting schemes or ensembles) which maximize their combined strengths for finding the right mix of linear and non-linear formula. Regardless of whether choosing just one or multiple algorithm(s), most benchmarks have improved performance when selecting better features. Feature selection here refers

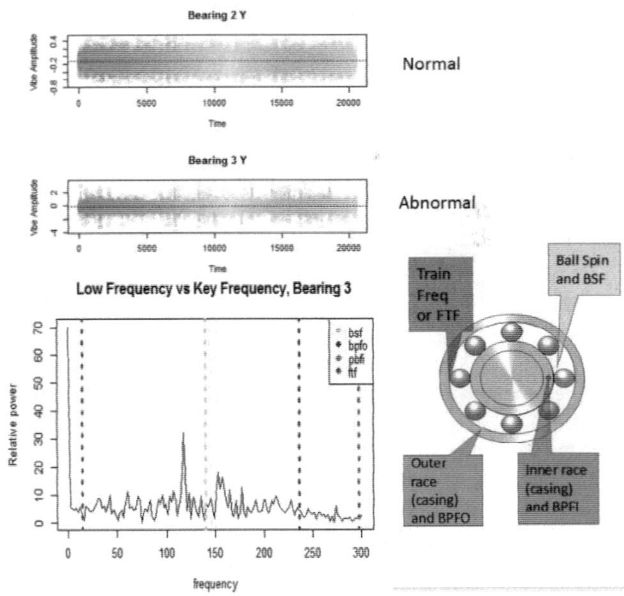

Figure 1 Spectral Power Not Evident in Failing Bearings (3) Near Key Ball Pass or Spin Frequencies

---

Presented at the AHS Airworthiness and HUMS Technical Meeting, Huntsville, Alabama, USA, February 21-22, 2018.

to the combination of states that accompany a given fault. For example, a single vibration spectrum will have many higher moments that may help understand the distribution tail or rarer signatures.

Table 1 Bearing Fault Features of Interest

Features	Description	Type
SD	Standard Deviation	Statistical
Skew	Skew	Statistical
Kurt	Kurtosis	Statistical
RMS	Root Mean Square	Statistical
Crest	Crest Factor	Statistical
Mean	Mean	Statistical
Qu.1	First Quantile	Statistical
FTF	Fundamental Train Frequency	Domain
BPFI	Ball Pass Frequency of Inner Ring	Domain
BPFO	Ball Pass Frequency of Outer Ring	Domain
BSF	Ball Spin Frequency	Domain
VHF.pow	Very High Frequency Spectral Power (6 kHz+)	Domain
HF.pow	High Frequency Spectral Power (2.6-6 kHz)	Domain
MF.pow	Mid Frequency Spectral Power (1.25 kHz -2.6 kHz)	Domain
LF.pow	Low Frequency Spectral Power (Forcing Frequency < 1.25 kHz)	Domain

The approach of the current research is to compare and contrast many statistical features and algorithms to see if fault prediction improves. The method is necessarily experimental. In many ways the ideal approach to condition based maintenance might be to input as many features as available, not just the dominant ones, and then let the algorithms converge by winnowing away the less important features from labeled training examples as done in Ref. 3. Forecasting mechanical faults from raw vibration signals requires spectral feature extraction and extensive algorithmic training. When fault signatures are applied as fielded thresholds, the results from a single classifier can suffer from variance (overfitting and dependence on the initial training set) and bias (poor overall testing accuracy). We examine this bias-variance tradeoff for a large sample set and compare diverse machine learning classifiers both individually and in combination. Methods of combining algorithms and features highlight their potential for providing new types of condition indicators and their weaknesses for stacking complexity. The method can be thought as treating algorithms and features as interchangeable building blocks that cascade into a combined data and learning pipeline.

METHODS

Novel aspects of this research center on a broad mix of algorithms to optimize fault detection. The approach first extracts classical spectral and time domain features from raw accelerometer data, then applies machine learning algorithms in a competitive way to capture essential tradeoffs between over-fitting and accuracy that plague traditional condition-based maintenance (CBM) metrics. The algorithm list takes advantage of diverse and often contrasting approaches to the detection problem. An example machine learning comparison is linear fits versus highly non-linear decision trees. The ensemble sum of many competing algorithms (and features) highlights the broader problem of handling noisy inputs and generalizing from what otherwise can prove difficult data assembly challenges.

Dataset Pre-Processing and Feature Extraction

The training (and test) fault dataset was selected from the open-sourced bearing experiments (originally provided by the Center for Intelligent Maintenance Systems (IMS, University of Cincinnati, Ref. 4) and hosted on the NASA Ames Prognostics Data Repository (Ref 5). In 7588 files consisting of one second vibration snapshots (20,480 points collected at 20 kHz) are the results of test-to-failure experiments on a four-bearing test rig with either one- or two-accelerometer(s) for x and y axes. Bearings 3 and 4 failed at the end of the test and 1 and 2 remained in service. Following the feature extraction protocol in Ref. 3, we apply domain knowledge about key bearing frequencies (FFT), statistical features (e.g. kurtosis) for all bearings in (x,y) components as summarized in Table 1. We did not include noise filters (e.g. wavelets), or automated factor combinations (e.g. log, sums). For 8624 cases the resulting dataset represents 29 features and a binary class assignment as either normal or abnormal (early damage, suspect, failure.roller, failure.b2, failure.inner, stage2). Thus 350

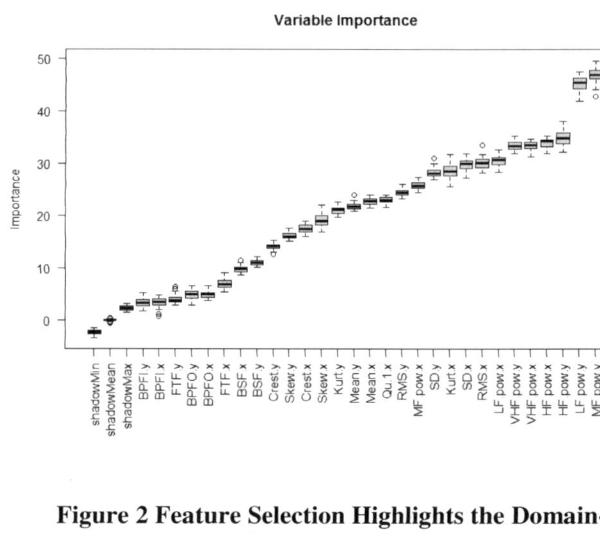

Figure 2 Feature Selection Highlights the Domain-Specific Factors

million raw vibration observations yield a much smaller sample to classify. As a nearly balanced (run-to-fail) dataset, the abnormal bearings represent 41.3% of all cases - much higher than expected from field observation but helpful to avoid overfit and imbalanced biases. Although the raw vibration data is large, feature selection winnows the training set to a small, relatively fast machine learning problem.

Description of Selected Features

We applied automated feature selection to rank order the importance of each factor. The approach assigns a top-down search based on Boruta in Ref 6. The top 7 most important detection factors for bearing abnormalities involve broad bands of mid, low and high frequency power. Consistent with Figure 1 visual inspection, the least important factors in Figure 2 are domain-specific frequencies related to ball and ring frequencies such as FTF, BPFI, BPFO, and BSF. This suggests that looking for unhealthy rotation frequencies in the bearing dataset may not classify abnormalities using simple peak-picks. The broad contribution of statistical features like skew and kurtosis suggests a multi-factor prediction algorithm might outperform single-factor thresholds for designing condition indicators.

Description of Classifier Suite

We implemented the classifier suite using the R algorithm libraries conveniently packaged in a larger wrapper library called caret. This machine learning wrapper described in Ref. 7 provides a common calling function, train, which simplifies large scale algorithm surveys. For bearing fault detection, the classifier suites include major families of linear multinomial and adaptive regressions; discriminant and principal component analysis; Bayesian; neural networks including deep and shallow layers; support vector machines; rule-based decision trees include ensembles such as boosting, bagging, stacking and random forests. We rank the algorithms' predictive accuracy and relative execution time. We validate performance using training and testing data partitioning, along with repeated cross-fold validation. We attempted no parameter tuning in favor of accepting model defaults. Models were excluded if they took longer than a few hours or failed to converge.

Family	Model	Accuracy	Time	Algorithm
Boosting	ADABOOST	0.9791	11.29	AdaBoost Classification Trees
Boosting	XGBLINEAR	0.9782	23.81	eXtreme Gradient Boosting
Decision trees	C5.0COST	0.9777	1.84	Cost-Sensitive C5.0
Decision trees	C5.0	0.9777	51.87	C5.0 Decision Trees and Rule-Based Models
Boosting	XGBTREE	0.9772	19.99	eXtreme Gradient Boosting
Boosting	DEEPBOOST	0.9708	17.38	Deep Boosting Ensemble Modeling
Bagging	TREEBAG	0.9693	6.57	Bagged CART
Bagging	BAGEARTHGCV	0.9659	2.31	Bagged MARS using gCV Pruning
Bagging	BAGEARTH	0.965	3.18	Bagged Multivariate adaptive regression splines
Random forests	CFOREST	0.9629	8.82	Conditional Inference Random Forest
Bagging	BAGFDA	0.9622	2.48	Bagged Flexible Discriminant Analysis
Bagging	BAGFDAGCV	0.962	1.35	Bagged FDA using gCV
Decision trees	J48	0.9592	6.82	C4.5-like Trees
Boosting	ADA	0.9537	1.01	Boosted Classification Trees
Boosting	BSTTREE	0.9506	34.62	Boosted Tree
Decision trees	CTREE	0.9462	2.00	Conditional Inference Tree
Discriminant Analysis	FDA	0.9456	4.30	Flexible Discriminant Analysis
Boosting	BLACKBOOST	0.9377	19.38	Boosted Tree
Decision trees	RPART1SE	0.9311	1.57	Classification and regression trees
Multivariate adaptive regression splines	GAMLOESS	0.9307	7.41	Generalized Additive Model using LOESS
Boosting	LOGITBOOST	0.9282	2.19	Boosted Logistic Regression
Generalized linear models	GLM	0.9079	1.13	Generalized Linear Model
Generalized linear models	BAYESGLM	0.9026	2.50	Bayesian Generalized Linear Model
Discriminant Analysis	AMDAI	0.8745	2.16	Adaptive Mixture Discriminant Analysis
Decision trees	RPART2	0.8035	1.53	Classification and regression trees
Boosting	GLMBOOST	0.7782	2.02	Boosted Generalized Linear Model
Decision trees	RPARTSCORE	0.7748	1.06	CART or Ordinal Responses
Decision trees	RPART	0.7748	1.72	Classification and regression trees
Decision trees	CTREE2	0.7734	2.12	Conditional Inference Tree
Discriminant Analysis	DDA	0.7628	4.33	Diagonal Discriminant
Decision trees	RPARTCOST	0.7607	1.95	Cost-Sensitive CART
Boosting	BSTSM	0.7215	6.96	Boosted Smoothing Spline
Neural networks	ELM	0.6998	1.38	Extreme Learning Machine
Partial least squares and principal component	GPLS	0.6822	2.59	Generalized Partial Least Squares
Boosting	BSTLM	0.6156	2.66	Boosted Linear Model

Figure 3 Algorithm Performance by Family, Method, Accuracy and Time

RESULTS

As shown in Fig. 3, we grouped and sorted classifier results by family, algorithm method, median accuracy and relative execution time (seconds). For detecting abnormal bearing vibration, simple rule-based partitions (like RPART) share some similarities with condition indicator thresholds and perform slightly better than chance given the distribution of normal-abnormal cases and rank-order in the bottom 20% of tested algorithms. The simple neural net model similarly underperforms in the bottom 8%. The poorest accuracy at 62% approximates little to no classification success for Boosted Linear Models. The general technique of boosting decision trees however scores the best for both accuracy (98%) and acceptable execution time (11 seconds). Tree-based learning as a group outperforms discriminant analysis, Bayesian and regression splines, neural nets, partial least squares, and linear models.

CONCLUSIONS

The strong performance of decision-tree ensembles with tailored convergence (such as boosting and bagging) is consistent with previous large-scale algorithm scoring in Refs 1-2. Applying decision trees for fault detection, particularly when combined with methods to prevent overfitting, offers intriguing alternatives to historically favored methods of spectral peak-picks, rule-based fault detection or legacy neural networks. As noted from a gap analysis in Ref. 8, the wider sharing of maintenance data offers the clearest path forward to discover better machine learning methods and improved condition-based decision-making. Future work will expand the classifier suite to include promising but under-represented families such as deep learning, random forests and extreme gradient-boosted models.

Author contact: David Noever david.noever@peopletec.com

ACKNOWLEDGMENTS

The author would like to thank the PeopleTec Technical Fellows program for encouragement and project assistance. This research benefited from support from U.S. Army Aviation Engineering Directorate, Redstone Arsenal, Alabama and the Special Operation Aviation Regiment (SOAR) Health Usage Maintenance (HUMS) Sustainment program, Ft. Campbell, KY.

REFERENCES

[1]Caruana, R., & Niculescu-Mizil, A. An empirical comparison of supervised learning algorithms. *ACM Proceedings of the 23rd International Conference On Machine Learning,* 2006, pp. 161-168.

[2]Fernández-Delgado, M., Cernadas, E., Barro, S., & Amorim, D. Do we need hundreds of classifiers to solve real world classification problems. *Journal of Machine Learning Research*, *15*(1), 2014, pp. 3133-3181.

[3] Catterson, V. Understanding data science, online tutorial, 2013, http://cowlet.org/2013/09/15/understanding-data-science-feature-extraction-with-r.html

[4] Qiu, H., Lee, J., Lin, J., & Yu, G. Wavelet filter-based weak signature detection method and its application on rolling element bearing prognostics. *Journal of Sound And Vibration*, *289*(4), 2006, pp. 1066-1090.

[5]Lee, J., Qiu, H., Yu, G., Lin,, J. and Rexnord Technical Services, IMS, University of Cincinnati. "Bearing Data Set", *NASA Ames Prognostics Data Repository* (http://ti.arc.nasa.gov/project/prognostic-data-repository), NASA Ames Research Center, Moffett Field, CA. 2007.

[6]Miron B., Kursa, W., Rudnicki, R.. Feature Selection with the Boruta Package. *Journal of Statistical Software, 36(11)*, 2010 p. 1-13.

[7]Kuhn, M. "Building Predictive Models in R Using the caret", *Journal of Statistical Software*, 28 (5), 2008, pp. 1-26

[8]Wade, D., Tucker, B., Davis, M., Knapp, D., Hasbroucq, S. Saporiti, M, Garrington, M., Rudy, A., "Joint Military and Commercial Rotorcraft Mechanical Diagnostics Gap Analysis," *American Helicopter Society 73th Annual Forum Proceedings*, Fort Worth, TX, May 2017

Self-Sensing Self-Diagnostic Structures for Next Generation Intelligent Aerospace Systems

Fotis Kopsaftopoulos
Assistant Professor
Mechanical, Aerospace, and Nuclear Engineering,
Rensselaer Polytechnic Institute
Troy, NY, USA

Fu-Kuo Chang
Professor
Department of Aeronautics & Astronautics, Stanford University
Stanford, CA, USA

ABSTRACT

The objective of this work is to outline a novel unified design, sensing, fabrication, and data-driven modeling and analysis approach for future self-sensing self-diagnostic intelligent aerospace structures with state sensing and awareness capabilities. The experimental assessment of is resented for an intelligent composite UAV wing with embedded distributed micro-sensor networks. The sensor networks consist of piezoelectric, strain, and temperature sensors in order to enable the data-driven modeling and interpretation and structural health monitoring of the wing under varying flight states and uncertainty. Piezoelectric sensors are employed in two modes: (i) passive mode to sense the ambient vibration of the wing in order to model and interpret the structural dynamic response and relate it to critical aerodynamic/aeroelastic phenomena such as stall and flutter; (ii) active mode, as both actuators and sensors to implement an active sensing acousto-ultrasound pitch-catch SHM approach. A novel modeling approach based on the recently introduced Vector-dependent Functionally Pooled (VFP) model structure is employed for the stochastic data-driven global modeling of the wing dynamics based on a series of wind tunnel experiments. In addition, the strain distribution is established under the considered flight states and critical areas of the flight envelope are identified. The obtained results demonstrate the successful integration of the micro-fabricated stretchable sensor networks with the composite materials of the wing, as well as the effectiveness of the stochastic "global" modeling and active sensing SHM approaches, proving their integration potential for the next generation of self-sensing self-diagnostic aerospace structures.

INTRODUCTION

The next generation of intelligent aerospace structures and aerial vehicles will be able to "feel," "think," and "react" in real time based on high-resolution state-sensing, awareness, and self-diagnostic capabilities. They will be able to sense and observe phenomena at unprecedented length and time scales allowing for improved performance, adaptability, autonomous operation, increased safety, reduced mission costs, and complete life-cycle monitoring and management. One of the main challenges of the current state-of-the-art research is the development of technologies that will lead to intelligent aerial vehicles able to (i) sense the external environment (temperature, air pressure, humidity, etc.) (Ref. 1), (ii) sense their flight and aeroelastic state (airspeed, angle of attack, aerodynamic loads, blade configuration and morphing, etc.) and internal structural condition (stresses, strains, damage) (Refs. 2–4), and (iii) effectively interpret the sensing data to achieve real-time state awareness, usage and health monitoring (Refs. 5–8).

Presented at the AHS Airworthiness and HUMS Technical Meeting, Huntsville, Alabama, USA, February 21-22, 2018.

Knowing the aerodynamic and structural loads, and health state of critical structural components, such as aircraft wings and rotorcraft blades, can be critical in the design of the next generation of self-sensing self-diagnostic structures. The integration of appropriate multifunctional composite materials with effective and robust monitoring and diagnostic approaches will enable high-resolution state sensing and awareness capabilities. Such self-sensing multifunctional materials have the potential to enable "intelligent" structural systems via the integration of HUMS and SHM systems in which a network of sensors is attached or embedded inside the composite structure. (Refs. 1, 6, 7, 9). The actual deployment of such approaches will lead to the improved performance, extended flight envelope, increased safety, and complete monitoring and life-cycle management of future aerial vehicles.

Existing usage and health monitoring approaches are based on the use of information obtained via various sensor technologies installed on critical structural and mechanical components along with corresponding diagnostic algorithms and data analytics. Using this information and corresponding processing methods, such systems can issue alerts and identify potentially dangerous states that can potentially lead to safety-critical events. The pilot or controller/flight management system may enable appropriate control actions in order to ensure system safety within the specified flight envelope that has been predetermined during the design phase. Within this context, the performance capabilities remain strictly confined within the flight envelope due to limited structural awareness, and oftentimes are rather conservative in order to ensure the safety of the system. On the other hand, the "self-sensing self-diagnostic" concept aims at the development of methods that leverage high levels of structural and aeroelastic awareness via the use of appropriate sensing, modeling and data-driven analysis techniques in order to exploit the full range of performance capabilities while ensuring the structural safety based on the "global" real-time monitoring of the vehicle.

The most critical challenge for the postulation of a complete and applicable self-diagnostic state-awareness framework for future aircraft and vertical flight systems is the effective modeling and interpretation of sensory data obtained under constantly changing dynamic environments, multiple flight states and varying structural health conditions. Evidently, all these different operating conditions have a significant impact on the structural dynamics and aeroelastic response. When it comes to the aeroelastic behavior, dynamic aeroelastic effects resulting from the interaction of the aerodynamic, elastic, and inertial forces require careful consideration throughout the design phase of the aircraft and pose a major safety-critical factor in the qualification of aircraft into service (Refs. 10–14). It is therefore evident that the flight states and operating conditions –characterized by several heterogeneous variables, such as airspeed, angle of attack (AoA), blade pitch, altitude, temperature, humidity, icing, and so on– may vary over time, and consequently affect the system dynamics and aeroelastic response. In such cases, the issue of the accurate data analysis, modeling, and interpretation under varying flight states remains a critical one that needs to be properly addressed.

In this work, the complete design, integration, and wind tunnel experimental assessment are presented for an intelligent composite high-lift low-drag wing with state-sensing and awareness capabilities. A composite UAV wing is outfitted with four stretchable sensor networks that are embedded inside the carbon/glass fiber composite layup. Each of the four sensor networks consists of 8 piezoelectric lead-zirconate titanate (PZT) sensors (disc PZTs $1/8$ in diameter), 6 strain gauges, and 23 resistive temperature detectors (RTDs). The fabricated wing can sense its structural state and surrounding environment during flight and interpret the sensing information to determine its actual operating state and flight configuration. Piezoelectric sensors are used to sense the vibration of the wing and identify the coupled airflow-structural dynamics. Strain gauges are used to determine the strain distribution of the wing and identify potential critical areas for the considered experimental conditions. Wind tunnel experiments were conducted for various angles of attacks and freesrteam airflow velocities for the investigation of a broad regime of flight conditions and structural states. The method of approach is divided into three tasks: (i) distributed multi-modal sensor networks for state sensing; (ii) structural design and fabrication via integrated sensor networks and composite materials; (iii) state awareness via statistical signal processing and stochastic identification techniques for data processing and interpretation; (iv) structural health monitoring via an acousto-ultrasound guided-wave-based active sensing SHM in order to detect and localize damage on the wing during its actual operation.

In this study, special emphasis is given to the data modeling and system identification via wind tunnel experiments under various flight (operating) conditions defined by multiple airspeeds and angles of attack (AoA). A novel modeling approach based on the recently introduced Vector-dependent Functionally Pooled (VFP) model structure (Refs. 7, 15, 16) is employed for the stochastic identification of the coupled airflow-structural dynamics for the complete range of the admissible flight states; that is all airspeeds and angles of attack of the wing that are considered in the wind tunnel experiments and may span the complete flight envelope. The unique characteristic of the VFP-based modeling approach is that it enables the analytical inclusion of the flight and/or structural state (see (Ref. 17)) on the coupled airflow-structural dynamics, as the model parameters depend functionally on the flight state.

DISTRIBUTED SENSOR NETWORKS AND WING INTEGRATION

Recently, micro-fabricated expandable sensor networks have been developed and deployed micro-scale sensors over macroscopic areas (Refs. 1, 9, 18–22). The component size is on the same order as an individual fiber in typical composite materials or scrim in film adhesives and small enough to be placed into a composite without structural modifications. These networks can be used in-situ, from the material fabrication throughout its service life, to monitor the cure process of composite materials, characterize material properties post-cure, and monitor the structural dynamics along with the health of the structure during its life cycle.

In this work four stretchable sensor networks with integrated distributed PZT, strain, and RTD sensors have been designed and fabricated (Refs. 1, 9, 18–21) so that they can be embedded inside the layup of the composite wing. Extensible wires connect the network nodes and serve as the signal communication channels. Before stretching, the network dimensions are 52.8 mm by 39.6 mm that after the stretching process expand to 140 mm by 105 mm yielding a 700% total surface area increase (Ref. 21). Each of the four sensor networks contains 8 piezoelectric sensors (round PZTs 3.175 mm in diameter), 6 strain gauges, and 24 RTDs. The total number of embedded sensors in the composite wing is 148.

The prototype wing was designed, constructed and tested at Stanford University. The designed wing is based on the cambered SG6043 high lift-to-drag ratio airfoil with a 0.86 m wing span, 0.235 m chord, and an aspect ratio of 7.32. In order to achieve the successful integration and fabrication of

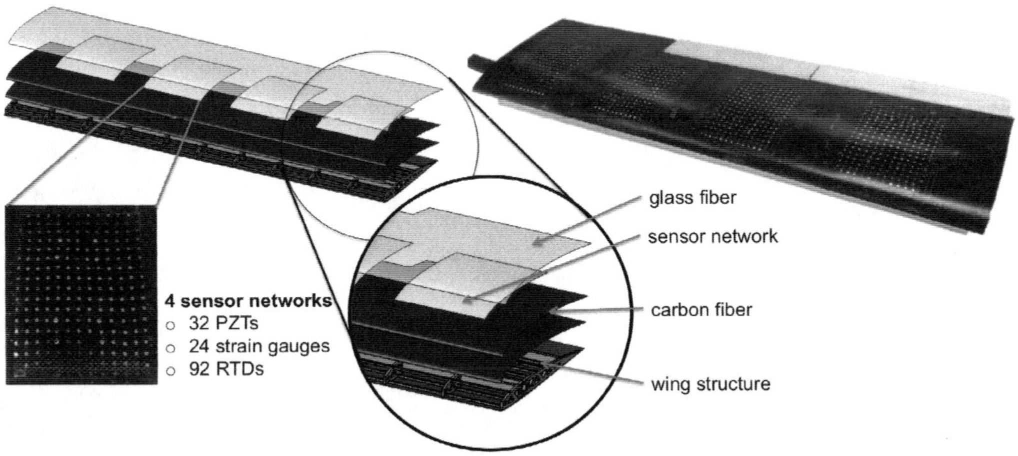

Fig. 1. The intelligent composite wing design with a total of 148 (32 piezoelectric, 24 strain gauges, and 92 RTDs) micro-sensors embedded in the composite layup.

the wing prototype, an appropriate network-material integration process had to be developed for embedding the microfabricated sensor networks inside the composite materials. The composite wing structure was manufactured based on carbon and glass laminated composites. The layup consists of carbon fiber (CF) plain weave fabric 1K T300 and glass fiber (GF) plain wave fabric 18 gr/m^2 infused with Araldite LY/HY5052 epoxy. The stacking sequence of the layers was [0° GF, 0° CF, 45° CF, 45° CF, 0° CF, 0° GF] (Figure 1). The four networks were embedded between the two top layers at 0° of the layup (near the wing surface) during the lamination process. The supporting wing structure consists of wooden (basswood) ribs and spars.

THE EXPERIMENTS

The wing was tested in the open-loop wind tunnel (WT) facility at Stanford University. The WT has a square test section of 0.84 m by 0.84 m (33 by 33 in) and can achieve continuous flow speeds up to approximately 28 m/s. A custom basis was designed and fabricated to support the wing and permit adjustments in the angle of attack (AoA). The wing was mounted horizontally inside the test section. Eight commercial strain gauges were attached on appropriate locations of the basis to measure the aerodynamic forces. The axis of rotation coincided approximately with the quarter of the wing chord. Figure 2 presents the composite wing with the corresponding locations of the PZTs and strain sensors. Table 1 presents the wing dimensions.

A series of wind tunnel experiments were conducted for various AoA and airspeeds. For each AoA, spanning the range from 0 to 18 degrees with an incremental step of 1 degree, data were sequentially collected for all velocities within the range 9 m/s to 22 m/s (incremental step of 1 m/s). The above procedure resulted to 266 different experiments covering the

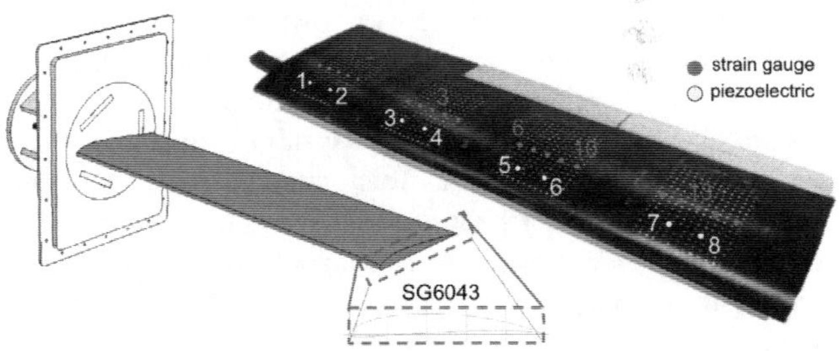

Fig. 2. The intelligent composite wing with the embedded sensor networks and the locations of the piezoelectric and strain sensors.

complete range of the considered flight states. The experimental conditions along with the Reynolds numbers are outlined in Table 2.

For each experiment the vibration and strain responses were recorded at different locations on the wing via the embedded piezoelectric sensors (initial sampling frequency $f_s = 1000$ Hz, initial signal bandwidth $0.1 - 500$ Hz) and strain gauges (sampling frequency $f_s = 100$ Hz, signal bandwidth DC100 Hz), respectively. The signals were recorded via a National Instruments X Series 6366 data acquisition module featuring eight 16-bit simultaneously sampled analog-to-digital channels. The strain signals were driven through a custom designed and built signal conditioning device into the data acquisition system. The total number of the sensor signals that were obtained was limited by the available number of chan-

Table 1. Wing dimensions.

Chord c	0.235 m
Span b	0.86 m
Area S	0.2 m^2
Aspect Ratio AR	7.32

Table 2. The conditions considered in the wind tunnel experiments.

Re ($\times 10^3$)	124	155	171	187	202	217	233	248	264	280	295	311	326	342
U_∞ (m/s)	9	10	11	12	13	14	15	16	17	18	19	20	21	22
Angle of attack: 0 – 18 degrees; Total number of experiments: 266														

Table 3. Data acquisition and signal pre-processing.

	Piezos	Strain gauges
Number of sensors:	8	15
Sampling frequency:	$f_s = 1000$ Hz	$f_s = 100$ Hz
Bandwidth:	$[0.1 - 500]$ Hz	$[DC - 50]$ Hz
Signal length:	$N = 90,000$ (90 s)	$N = 9,000$ (90 s)

nels of the data acquisition system. Table 3 presents the sensors, data acquisition, signal details.

STOCHASTIC GLOBAL MODELING

In this section the modeling of the coupled airflow-structural dynamics is addressed via the use of stochastic functionally pooled models, or more precisely Vector-dependent Functionally Pooled AutoRegressive (VFP-AR) models. These models are capable of representing the system dynamics for the complete range of flight and structural states based on data records obtained under a sample of these states. The problem is important in a number of practical applications and is tackled within a recently introduced Functional Pooling framework (Refs. 7, 15, 16, 23). This study focuses on the case of flight states characterized by two parameters, namely the airspeed and the angle of attack of the wing. For the case where the modeling of the structural state is modeled in terms of damage location and damage magnitude, the interested reader is referred to (Refs. 16, 17).

Classical system identification aims at deriving a model representing a system under a specific operating condition. Yet, in many cases, a system may operate under different conditions at different occasions (time periods), with the dynamics depending in a pseudo-static fashion on certain operating parameter(s) - also referred to as scheduling parameter(s). In such cases, given a number of data records from the system operating under a sample of different conditions, it is highly desirable to establish a single and global model, that, while compact (parsimonious), will be capable of accurately representing the dynamics under any considered condition.

Baseline Modeling under a Single Flight State

The modeling of the wing's dynamic response under a single flight state is an initial step performed in order to facilitate (in the sense of providing approximate model orders) the subsequent step of the global modeling under all the admissible flight states. A discrete-time model (or a vector model or an array of models in the case of several vibration response measurement locations) representing the system dynamics is obtained via standard identification procedures (Refs. 24, 25).

In this study a single response AutoRegressive (AR) model is used. An AR(n) model is of the form[1] (Ref. 24):

$$y[t] + \sum_{i=1}^{n} a_i \cdot y[t-i] = e[t] \qquad e[t] \sim \text{iid}\,\mathcal{N}\left(0, \sigma_e^2\right) \qquad (1)$$

with t designating the normalized discrete time ($t = 1,2,3,\ldots$ with absolute time being $(t-1)T_s$, where T_s stands for the sampling period), $y[t]$ the measured vibration response signals as generated by the piezoelectric sensors of the wing, n the AR order, and $e[t]$ the stochastic model residual (one-step-ahead prediction error) sequence, that is a white (serially uncorrelated), Gaussian, zero mean with variance σ_e^2 sequence. The symbol $\mathcal{N}(\cdot,\cdot)$ designates Gaussian distribution with the indicated mean and variance, and iid stands for identically independently distributed.

The model is parameterized in terms of the parameter vector $\bar{\theta} = [a_1 \ \ldots \ a_n \,\vdots\, \sigma_e^2]^T$ to be estimated from the measured signals (Ref. 24). Model estimation may be achieved based on minimization of the Ordinary Least Squares (OLS) or the Weighted Least Squares (WLS) criteria (Ref. 24). The modeling procedure involves the successive fitting of AR(n) models for increasing order n until an adequate model is selected (Ref. 26). Model order selection is based on the Bayesian Information Criterion (BIC) and the residual sum of squares normalized by the series sum of squares (RSS/SSS). Final model validation is based on formal verification of the residual (one-step-ahead prediction error) sequence uncorrelatedness (whiteness) hypothesis (Ref. 24, pp. 512-513).

Global Modeling under Multiple Flight States

The VFP-AR representation allows for complete and precise modeling of the global wing dynamics under multiple operating conditions defined by varying airspeed and angle of attack. The VFP model parameters and residual series covariance depend functionally on the airspeed and AoA, while the corresponding interrelations and statistical dependencies between the different flight states are taken into account.

The global modeling of the composite wing via a VFP-AR model involves consideration of all the admissible airspeeds and AoA that define the flight envelope of the wing. A total of $M_1 \times M_2$ experiments is performed (in this case via wind tunnel experiments; alternatively, analytical models and simulations can be used), with M_1 and M_2 designating the number of experiments under the various airspeeds and AoA, respectively. Each experiment is characterized by a specific

[1]Lower case/capital bold face symbols designate vector/matrix quantities, respectively.

airspeed k^1 and a specific AoA k^2, with the complete series covering the required range of each variable, say $[k^1_{min}, k^1_{max}]$ and $[k^2_{min}, k^2_{max}]$, via the discretizations $\{k^1_1, k^1_2, \ldots, k^1_{M_1}\}$ and $\{k^2_1, k^2_2, \ldots, k^2_{M_2}\}$. For the identification of a global VFP model the flight state vector $\mathbf{k}$ containing the airspeed and AoA scalar components, is defined as:

$$\mathbf{k} = [k^1_i \; k^2_j]^T \Longleftrightarrow k_{i,j}, \quad i = 1, \ldots, M_1, \quad j = 1, \ldots, M_2 \quad (2)$$

with $k_{i,j}$ designating the flight state of the wing corresponding to the i-th airspeed and the j-th angle of attack. This procedure yields a pool of response signals (each of length N):

$$x_{\mathbf{k}}[t], y_{\mathbf{k}}[t] \quad \text{with} \quad t = 1, \ldots, N, \quad k^1 \in \{k^1_1, \ldots, k^1_{M_1}\},$$

$$\text{and} \quad k^2 \in \{k^2_1, \ldots, k^2_{M_2}\}. \quad (3)$$

The VFP-AR model is of the following form (Refs. 7, 15):

$$y_{\mathbf{k}}[t] + \sum_{i=1}^{n} a_i(\mathbf{k}) \cdot y_{\mathbf{k}}[t-i] = e_{\mathbf{k}}[t] \quad (4)$$

$$e_{\mathbf{k}}[t] \sim \text{iid} \mathcal{N}\left(0, \sigma_e^2(\mathbf{k})\right) \qquad \mathbf{k} \in \mathbb{R}^2 \quad (5)$$

$$a_i(\mathbf{k}) = \sum_{j=1}^{p} a_{i,j} \cdot G_j(\mathbf{k}) \quad (6)$$

$$E\{e_{k_{i,j}}[t] \cdot e_{k_{m,n}}[t-\tau]\} = \gamma_e[k_{i,j}, k_{m,n}] \cdot \delta[\tau] \quad (7)$$

with n designating the AR order, $y_{\mathbf{k}}[t]$ the sensor's response signal, and $e_{\mathbf{k}}[t]$ the model's residual (one-step-ahead prediction error) sequence, that is a white (serially uncorrelated) zero mean sequence with variance $\sigma_e^2(\mathbf{k})$. This may potentially be cross-correlated with its counterparts corresponding to different experiments (different $\mathbf{k}$'s). The symbol $E\{\cdot\}$ designates statistical expectation, $\delta[\tau]$ the Kronecker delta (equal to unity for $\tau = 0$ and equal to zero for $\tau \neq 0$), $\mathcal{N}(\cdot,\cdot)$ Gaussian distribution with the indicated mean and variance, and iid stands for identically independently distributed.

As (6) indicates, the AR parameters $a_i(\mathbf{k})$ are modeled as explicit functions of the flight vector $\mathbf{k}$ (which contains the airspeed and AoA components) by belonging to p-dimensional functional subspace spanned by the mutually independent basis functions $G_1(\mathbf{k}), G_2(\mathbf{k}), \ldots, G_p(\mathbf{k})$ (*functional basis*). The functional basis consists of polynomials of two variables (bivariate) obtained as tensor products from their corresponding univariate polynomials (Chebyshev, Legendre, Jacobi, and other families (Refs. 15, 16)). The constants $a_{i,j}$ designate the AR coefficients of projection.

The VFP-AR model of (4)–(7) is parameterized in terms of the parameter vector to be estimated from the measured signals:

$$\bar{\theta} = [\, a_{1,1} \; a_{1,2} \; \ldots \; a_{i,j} \vdots \sigma_e^2(\mathbf{k}) \,]^T \quad \forall \mathbf{k} \quad (8)$$

and may be written in linear regression form as:

$$y_{\mathbf{k}}[t] = [\varphi^T_{\mathbf{k}}[t] \otimes g^T(\mathbf{k})] \cdot \theta + e_{\mathbf{k}}[t] = \phi^T_{\mathbf{k}}[t] \cdot \theta + e_{\mathbf{k}}[t] \quad (9)$$

with:

$$\varphi_{\mathbf{k}}[t] := \Big[-y_{\mathbf{k}}[t-1] \ldots -y_{\mathbf{k}}[t-n]\Big]^T_{[n\times 1]} \quad (10)$$

$$g(\mathbf{k}) := \Big[G_1(\mathbf{k}) \ldots G_p(\mathbf{k})\Big]^T_{[p\times 1]} \quad (11)$$

$$\theta := \Big[a_{1,1} \; a_{1,2} \; \ldots \; a_{n,p}\Big]^T_{[(np\times 1]} \quad (12)$$

and ^{T} designating transposition and $\otimes$ Kronecker product (Ref. 27, Chap. 7).

Pooling together the expressions (9) of the VFP-AR model corresponding to all flight vectors $\mathbf{k}$ $(k_{1,1}, k_{1,2}, \ldots, k_{M_1,M_2})$ considered in the experiments (cross-sectional pooling) yields:

$$\begin{bmatrix} y_{k_{1,1}}[t] \\ \vdots \\ y_{k_{M_1,M_2}}[t] \end{bmatrix} = \begin{bmatrix} \phi^T_{k_{1,1}}[t] \\ \vdots \\ \phi^T_{k_{M_1,M_2}}[t] \end{bmatrix} \cdot \theta + \begin{bmatrix} e_{k_{1,1}}[t] \\ \vdots \\ e_{k_{M_1,M_2}}[t] \end{bmatrix}. \quad (13)$$

Then, following substitution of the data for $t = 1, \ldots, N$ the following expression is obtained:

$$y = \Phi \cdot \theta + e \quad (14)$$

with

$$y := \begin{bmatrix} y[1] \\ \vdots \\ y[N] \end{bmatrix}, \quad \Phi := \begin{bmatrix} \Phi[1] \\ \vdots \\ \Phi[N] \end{bmatrix}, \quad e := \begin{bmatrix} e[1] \\ \vdots \\ e[N] \end{bmatrix}. \quad (15)$$

Using the above linear regression framework, the projection coefficient vector θ can be efficiently estimated via minimization of the Weighted Least Squares (WLS) criterion:

$$J^{\text{WLS}} := \frac{1}{N} \sum_{t=1}^{N} e^T[t] \Gamma^{-1}_{e[t]} e[t] = \frac{1}{N} e^T \Gamma_e^{-1} e \quad (16)$$

which leads to the *Weighted Least Squares (WLS)* estimator:

$$\widehat{\theta}^{\text{WLS}} = [\Phi^T \Gamma_e^{-1} \Phi]^{-1} [\Phi^T \Gamma_e^{-1} y]. \quad (17)$$

In these expressions $\Gamma_e = E\{ee^T\}$ ($\Gamma_e = \Gamma_{e[t]} \otimes I_N$, with I_N designating the $N \times N$ unity matrix) designates the residual covariance matrix, which is practically unavailable. Nevertheless, it may be consistently estimated by applying (in an initial step) Ordinary Least Squares (details in (Ref. 15)). Once $\widehat{\theta}^{\text{WLS}}$ has been obtained, the final residual variance and residual covariance matrix estimates are obtained as:

$$\widehat{\sigma}_e^2(\mathbf{k}, \widehat{\theta}^{\text{WLS}}) = \frac{1}{N} \sum_{t=1}^{N} e^2_{\mathbf{k}}[t, \widehat{\theta}^{\text{WLS}}],$$

$$\widehat{\Gamma}_{e[t]} = \frac{1}{N} \sum_{t=1}^{N} e[t, \widehat{\theta}^{\text{WLS}}] e^T[t, \widehat{\theta}^{\text{WLS}}]. \quad (18)$$

The estimator $\widehat{\theta}^{\text{WLS}}$ may, under mild conditions, be shown be asymptotically Gaussian distributed with mean coincidi with the true parameter vector θ^o and covariance matrix P_θ

$$\sqrt{N}(\widehat{\theta}_N - \theta^o) \sim \mathcal{N}(0,P_\theta) \quad (N \longrightarrow \infty) \qquad (1$$

based on which interval estimates of the true parameter vect may be constructed (Ref. 15).

The problem of VFP-AR model structure selection (structu estimation) for a given basis function family (such as Cheb shev, Legendre, and so on), that is model order determinati for the AR polynomial and determination of their correspon ing functional subspaces, is referred to as the model identi cation problem. Usually, the AR model order is initially s lected via customary model order selection techniques (BI RSS, frequency stabilization diagrams) (Ref. 20), whereas t functional subspace dimensionality is selected via a Genetic Algorithm (GA) procedure (Ref. 15). Initially, the maximum functional subspace dimensionality is selected, which defines the search space of the functional subspace estimation subproblem. The determination of the exact subspace dimensionality is achieved via the use of GAs based on minimization of the BIC with respect to the candidate basis functions. In the current study, the estimation of the functional subspace dimensionality was achieved via the use of the BIC criterion for increasing functional subspace dimensionality.

EXPERIMENTAL RESULTS

Non-parametric analysis

Figure 3 presents indicative wind tunnel signals obtained from piezoelectric sensor 2 under various AoA and freestream velocities of $U_\infty = 11$ m/s (top subplot) and $U_\infty = 15$ m/s (bottom subplot). Observe the stochastic (random) nature of these signals, which is due to the wind tunnel airflow actuation and the aeroelastic response of the wing. In addition, it is evident that for higher AoA and as the wing approaches stall, the signal amplitude (voltage) increases. In the case of $U_\infty = 11$ m/s (top subplot) in Figure 3, the maximum signal amplitude for AoA of 13 and 15 degrees seems to be similar as there is no evident further increase. For this freestream velocity and based on the airfoil properties and CFD analysis, stall occurs at an AoA of approximately 13 degrees. In the case of $U_\infty = 15$ m/s (bottom subplot) in Figure 3, stall occurs at approximately 15 degrees, and it may be readily observed that there is an obvious increase in the signal amplitude from 13 to 15 degrees AoA.

Non-parametric identification is based on 90,000 (90 s) sample-long response signals obtained from the embedded piezoelectric sensors (see Table 3). A 5096 sample-long Hamming data window (frequency resolution $\Delta f = 0.24$ Hz) with 90% overlap is used for the Welch-based spectral estimation (MATLAB[2] function *pwelch.m*).

[2] In this work Matlab version R2015b has been used.

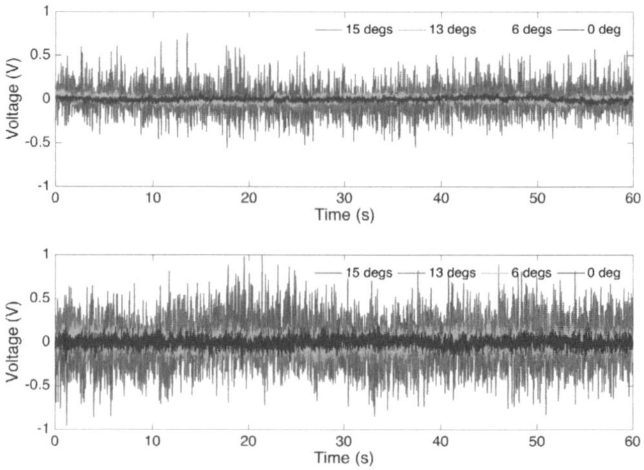

Fig. 3. Indicative signals obtained from piezoelectric sensor 1 under various angles of attack: (a) freestream velocity $U_\infty = 11$ m/s (top subplot) and (b) freestream velocity $U_\infty = 17$ m/s (bottom subplot).

Figure 4 presents indicative non-parametric power spectral density (PSD) Welch-based estimates obtained from piezoelectric sensor 1 for increasing airspeed and a constant AoA of 0 degrees within the $[0.1 - 25]$ Hz frequency range (left subplot). Notice that as the airspeed increases, the PSD amplitude in the lower frequency range slightly increases as well. In this case, it is expected that as the airspeed increases for a constant AoA the wing will approach flutter which will be triggered by the coupling of aeroelastic modes. In this case the coupling occurs in the $[0.5 - 15]$ Hz frequency range. By carefully observing Figure 4 it may be seen that the frequency at approximately 5 Hz increases with increasing airspeed and approaches the frequency at approximately 9 Hz, thus providing an indication of incipient flutter. This observation will be clarified by the global parametric modeling results.

Similarly, Figure 5 presents indicative non-parametric power spectral density (PSD) Welch-based estimates of the piezoelectric response signals obtained from sensor 1 for increasing AoA and freestream velocity $U_\infty = 15$ m/s (right subplot). Notice that as the AoA increases the PSD amplitude in the lower frequency range of $[0.1 - 12]$ Hz significantly increases as well. More specifically, as the AoA approaches the critical stall range of $[13 - 15]$ degrees, the low frequency vibrations become dominant and thus indicating the proximity to the stall of the wing. From this Figure it is evident that by monitoring the identified lower frequency bandwidths that are sensitive to increasing AoA it is possible to obtain a strong indication of stall. All the embedded piezoelectric sensors of the wing exhibit a similar performance, but for the sake of brevity the results are presently omitted

Strain Distribution

Indicative results obtained from strain gauges 5 and 14 are presented in Figure 6 and Figure 7, respectively. The recorded

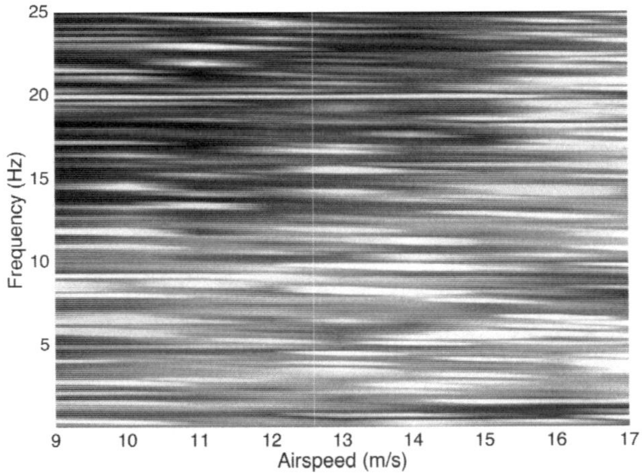

Fig. 4. Indicative non-parametric Welch-based power spectral density (PSD) estimates for piezoelectric sensor 1 versus varying airspeed with set AoA of 1 degree.

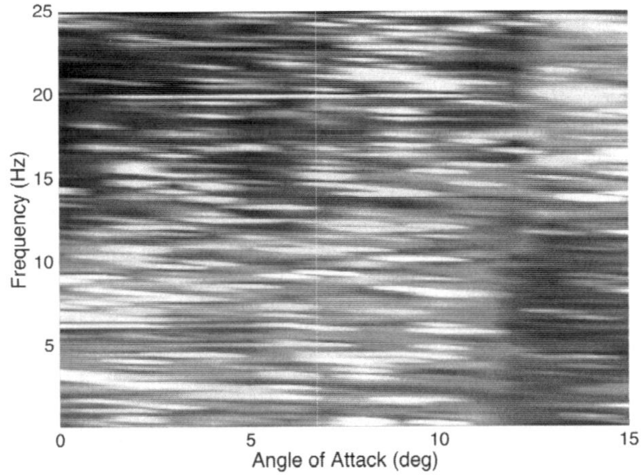

Fig. 5. Indicative non-parametric Welch-based power spectral density (PSD) estimates for piezoelectric sensor 1 versus varying AoA for set airspeed $U_\infty = 13$ m/s.

strain is plotted for increasing AoA and airspeed, hence the strain mapping for the specific location on the wing may be obtained for the considered experimental conditions. As the AoA and airspeed increase the applied strain also increases, with the maximum strain appearing, as expected, in the top right corner of the figures (for the maximum airspeed and AoA). The strain shown in these plots is estimated as the mean value of the acquired signal (9000 samples; 90 s). The maximum strain is observed for angles higher than 8 degrees and velocities higher than 16 m/s, resulting in a critical, with respect to strain, flying area of the wing that for the considered experimental conditions (or –in the more general concept– flight envelope) is located in the top right corner of the figures. Finally, the recorded strain is higher for gauge 5 than gauge 13. This is due to the fact that strain gauge 5 is closer to the root of the wing and thus it is expected that the quasi-static strain in that location should be higher than the strain closer to the wing tip.

Parametric analysis via global models

For the parametric identification, conventional autoregressive (AR) time-series models representing the wing]s structural dynamics are obtained through standard identification procedures (Refs. 24, 25) based on the collected piezoelectric response signals (MATLAB function *arx.m*). The response signal bandwidth is selected as $0.1-30$ Hz after the initial signals were low-pass filtered (Chebyshev Type II) and sub-sampled to a resulting sampling frequency $f_s = 60$ Hz (initial sampling frequency was 1000 Hz). Each signal resulted in a length of $N = 4{,}000$ samples (20 s) and was subsequently sample mean corrected.

The modeling strategy consists of the successive fitting of AR(n) models (with n designating the AR order) until a suitable model is selected. Model parameter estimation is achieved by minimizing a quadratic prediction error (PE) criterion leading to a least squares (LS) estimator (Ref. 24, p. 206). Model order selection, which is crucial for successful identification, may be based on a combination of tools, including the Bayesian information criterion (BIC) , which is a statistical criterion that penalizes model complexity (order) as a counteraction to a decreasing quality criterion (Ref. 24, pp. 505–507), monitoring of the RSS/SSS (residual sum of squares/ signal sum of squares) criterion, monitoring of the residual autocorrelation function (MATLAB function autocorr.m) (Ref. 24, p. 512), and use of "stabilization diagrams" which depict the estimated modal parameters (usually frequencies) as a function of increasing model order (Refs. 24, 25).

For piezoelectric sensor 1, this leads to an AR(18) model for a collected data set corresponding to an airspeed of 11 m/s and an angle of attack of 3 degrees. This model is used as reference and for providing approximate orders for the identification of the global VFP-AR models of the next section. For the sake of brevity, in the following sections indicative results from sensor 1 only will be presented.

The global modeling of the composite wing is based on signals obtained from a total of $M_1 \times M_2 = 144$ experiments. Airspeeds up to 17 m/s and AoA up to 15 degrees were considered for the Vector-dependent Functionally Pooled (VFP) based modeling procedure. The airspeed and AoA increments are $\Delta k^1 = 1$ m/s and $\Delta k^2 = 1$ degree, respectively, covering the corresponding intervals of $[9, 17]$ m/s and $[0, 15]$ degrees.

Model order selection starts with the orders selected for the conventional AR models representing the wing structure for a constant indicative experimental condition. The final model orders being presently selected are based on the BIC criterion (Ref. 26) and model validation techniques, such as checking the whiteness (uncorrelatedness) and the normality of the model residuals (MATLAB functions *acf.m* and *normplot.m*, respectively) (Refs. 24, 26). The functional subspaces are selected via a similar BIC-based process. The functional subspace consists of 21 Chebyshev Type II bivariate polynomial

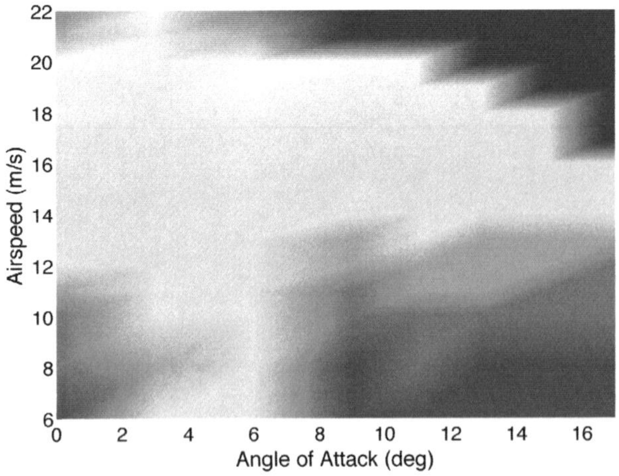

Fig. 6. Indicative strain distribution obtained from strain gauge 5 for the considered angles of attack and freestream velocities.

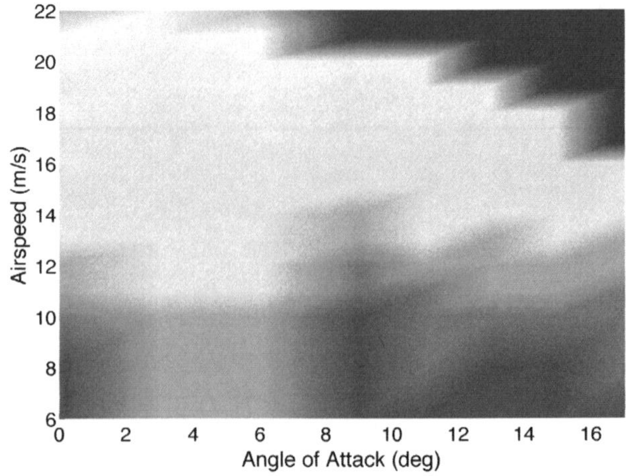

Fig. 7. Indicative strain distribution obtained from strain gauge 13 for the considered angles of attack and freestream velocities.

basis functions (Refs. 15–17).

The final identified global model is a VFP-AR$(72)_{24}$. The functional subspace consists of the first $p = 24$ shifted Chebyshev Type II 2-dimensional polynomials.

In this study, the identified VFP model is of significantly lower order than the corresponding results presented in (Ref. 7). This was to investigate a model that focuses on the lower frequencies (< 30 Hz) that are dominant in the structural response of the composite wing. This also reduces potential over-parametrization and significantly reduces the computational burden allowing for potential close to real-time model estimation.

Indicative VFP-model-based frequency response function (FRF) magnitude results obtained from the VFP-AR$(18)_{21}$ global model are depicted as functions of frequency and airspeed for set AoA 0 degrees in Figure 8. The frequency resolution is 0.01 Hz, while the airspeed resolution is 0.1 m/s. Observe how the wing mode at 4.5 Hz for airspeed 9 m/s gradually increases with the increasing airspeed until completely coupled with the mode at 9 Hz at approximately 16 m/s. This behavior of the aeroelastic modes of the wing, as identified by the VFP-AR model, corresponds to the generation of dynamic flutter. It may be readily observed that the results of Figure 8 are, as expected, extremely accurate when compared to the corresponding non-parametric Welch-based analysis of Figure 4. It is also worth mentioning that the non-parametric results of Figure 4 have been obtained using a significantly longer signal of 90 seconds, whereas the VFP-based parametric results are based on 20-second-long signals.

The VFP-model-based FRF magnitude curves obtained via the VFP-AR$(18)_{21}$ global model are depicted as functions of frequency and AoA in the right plot in Figure 9. The frequency resolution is 0.01 Hz, while the AoA resolution is 0.1 degrees. The airspeed of 15 m/s is very close to the occurrence of flutter and the wing exhibits two distinct aeroelastic modes within the $[5-10]$ Hz range. However, observe that for an AoA of 0 degrees, the modes are very close and almost coincide, which is in agreement with Figure 8 that presents the model-based FRF for set AoA of 0 degrees versus increasing airspeed. By observing the frequency evolution versus the AoA it may be assessed that the amplitude of the VFP-based FRF magnitude increases for lower frequencies (< 15 HZ) with increasing AoA as the wing approaches stall. More specifically, the FRF magnitude exhibits a sharp increase for AoA higher than 13 degrees in which stall occurs.

By comparing the VFP-based parametric FRF magnitudes with the corresponding non-parametric Welch-based spectral estimates of Figure 5 it may be concluded that high accuracy is achieved by the global modeling approach which also employs a significantly shorter signal length (see Table 3).

ACTIVE SENSING SHM

SHM technologies, oftentimes in combination with appropriate NDE and/or HUMS, along with sophisticated data management systems and life-prediction models, address the formulation of the required steps for the transition to condition-based maintenance (CBM) and complete structural awareness for safety assurance. In this section, the application of an acousto-ultrasound active sensing SHM approach is presented for the intelligent composite wing. Apart from the passive sensing of the ambient vibration (aeroelastic) response of the wing via the embedded pizeoelectric sensor network, the same network can be used in an active sensing mode to perform pulse-echo or pitch-catch guided-wave-based diagnostic SHM. Piezoelectric sensors-actuators can be used both as transmitters and receivers for monitoring local defects by injecting controlled diagnostic signals into structures and can potentially interrogate large structural areas. In principle, an active system allows damage to be interrogated by injecting controlled diagnostic signals (i.e. guided Lamb waves) into the structure. With the known inputs, the changes in local

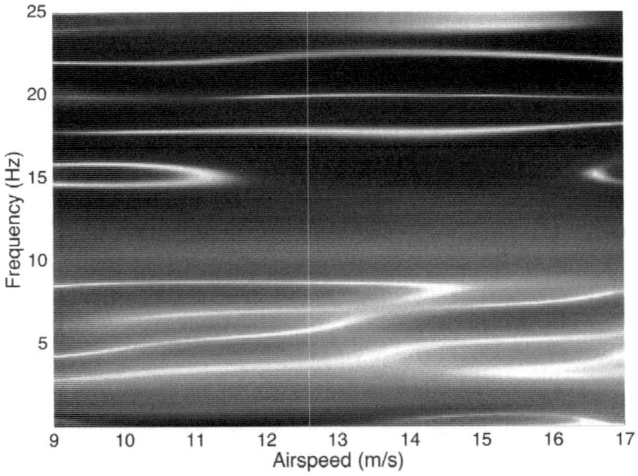

Fig. 8. Indicative parametric FRF magnitude results based on the VFP-AR$(18)_{21}$ global model as a function of frequency and airspeed for set AoA of 0 degrees.

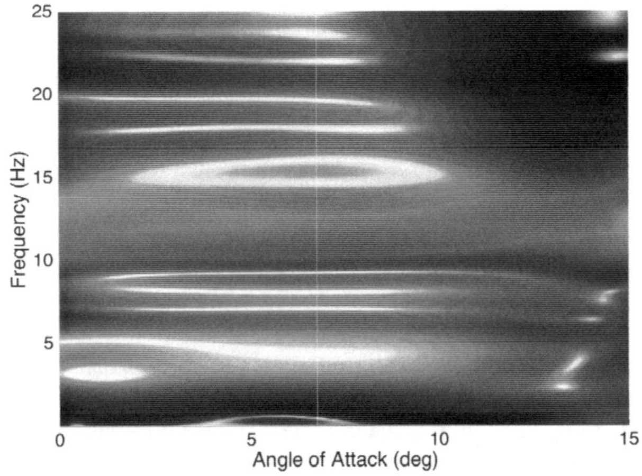

Fig. 9. Indicative parametric FRF magnitude results based on the VFP-AR$(18)_{21}$ global model as a function of frequency and AoA for set airspeed of 15 m/s.

sensor measurements are associated with the introduction of damage in the structure.

The use of distributed piezoelectric sensors embedded inside the wing allows for the development of in-situ, on-demand, diagnostic approaches that tackle damage detection, localization, and quantification; the present study focuses on the detection and localization tasks. The distributed network is expected to increase the damage localization and quantification accuracy and robustness, that is the topic of on-going work by the authors and co-workers (Refs. 28, 29). Increasing the damage quantification robustness of active sensing ultrasound approaches for hotspot monitoring can be achieved via the use of multiple sensors, and thus wave propagation paths (Ref. 29). As the number of sensors and propagation paths increases, the uncertainty in the damage index (as defined in (Ref. 28)) will decrease leading to robust damage quantification. Theoretically, it is assumed that the damage index value will converge to its "real" value as the number of sensors/paths increases, as the addition of new paths will not contribute to any useful information (Ref. 29).

Figure 10 presents indicative structural health monitoring results based on a series of experiments performed on the composite wing under controlled laboratory environment and during the wind tunnel experiments. Ultrasound guided 5-peak tone-burst waves were generated from the embedded piezoelectric transducers acting as actuators for a number of center frequencies, spanning the range of 100 – 700 kHz. The ultrasound waves were recorded by piezoelectric transducers acting as sensors and mounted on three sensing topologies on the wing. These topologies were defined in terms of a hotspot monitoring scenario, such as that specific locations on the wing are considered to be prone to the appearance of damage. In this case, the three topologies cover the complete area of the wing and employ 19 piezoelectric sensors embedded in the composite layup. The location of the sensors and the corresponding topologies are shown in Figure 10.

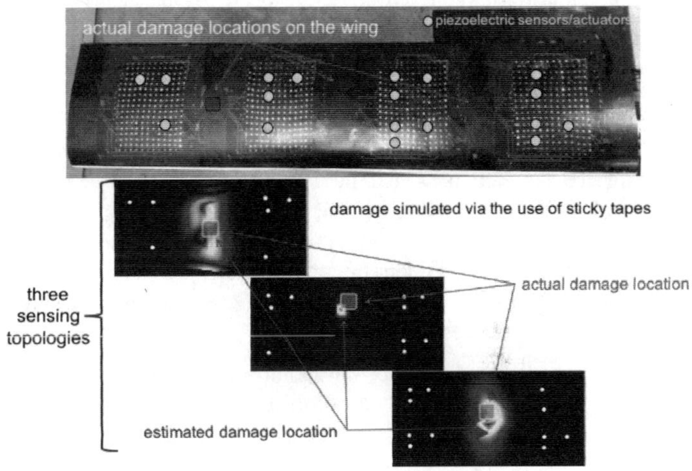

Fig. 10. Indicative structural health monitoring results based on the active-sensing acousto-ultrasound approach (Ref. 28) for a center frequency of 250 kHz.

Simulated damage was introduced in various locations on the wing by attaching lightweight sticky tapes. The active sensing approach employed in this study is based on the damage index (DI) defined in the recent work of (Ref. 28). The indicative results of Figure 10 for a center wave frequency of 250 kHz indicate that the method is able to detect and localize the existence of artificial damage. The actual damage location is indicated via the red rectangular areas, while the estimated damage location is indicated by the diagnostic image (for more details the interested reader is referred to (Ref. 5)).

In addition, the method was applied during the wind tunnel experiments and is the topic of on-going work to be presented in a subsequent article.

CONCLUSIONS

The objective of this work was to outline the main design, fabrication, and data modeling and analysis challenges of a self-sensing self-diagnostic intelligent composite UAV wing with state sensing and awareness capabilities. The embedded bio-inspired stretchable sensor networks consisting of piezoelectric, strain gauges, and RTDs enabled the data-driven identification of the coupled structural and aerodynamic properties under varying flight states and uncertainty. Piezoelectric sensors were used in two modes: (i) passive mode to sense the ambient vibration of the wing in order to identify the coupled airflow-structural dynamics and related them to critical aerodynamic/aeroelastic phenomena such as stall and flutter; (ii) active mode, as both actuators and sensors to implement an active sensing acousto-ultrasound pitch-catch SHM approach.

Special emphasis was given to the wind tunnel experimental assessment under various flight states defined by multiple airspeeds and angles of attack. A novel modeling approach based on the recently introduced Vector-dependent Functionally Pooled (VFP) model structure was employed for the stochastic data-driven modeling of the global aeroelastic vibration response of the wing. In addition, the strain distribution was established under the considered flight states and critical areas –as defined by increased strain signatures– of the flight envelope were identified. The obtained results demonstrated the successful integration of the micro-fabricated stretchable sensor networks with the composite materials of the wing, as well as the effectiveness of the stochastic VFP-based "global" modeling and active sensing SHM approaches, proving their integration potential for the next generation of self-sensing self-diagnostic aerial vehicles and aerospace structures.

Author contact: Fotis Kopsaftopoulos kopsaf@rpi.edu and Fu-Kuo Chang fkchang@stanford.edu

ACKNOWLEDGMENTS

This experiments and preliminary analysis for this work were supported by the U.S. Air Force Office of Scientific Research (AFOSR) program "Avian-Inspired Multifunctional Morphing Vehicles" under grant FA9550-16-1-0087 with Program Manager Byung-Lip ("Les") Lee. The author would like to thank Mr. Raphael Nardari and Dr. Yu-Hung Li at Stanford University for their contribution in the design and fabrication of the composite wing and sensor networks.

REFERENCES

[1]Salowitz, N., Guo, Z., Roy, S., Nardari, R., Li, Y.-H., Kim, S.-J., Kopsaftopoulos, F., and Chang, F.-K., "Recent advancements and vision toward stretchable bio-inspired networks for intelligent structures," *Structural Health Monitoring*, Vol. 13, (6), 2014, pp. 609–620.

[2]Kopsaftopoulos, F., Nardari, R., Li, Y.-H., and Chang, F.-K., "Experimental identification of structural dynamics and aeroelastic properties of a self-sensing smart composite wing," Structural Health Monitoring 2015: System Reliability for Verification and Implementation – Proceedings of the 10th International Workshop on Structural Health Monitoring (IWSHM 2015), edited by F.-K. Chang and F. Kopsaftopoulos, September 2015.

[3]Suryakumar, V. S., Babbar, Y., Strganac, T. W., and Mangalam, A. S., "Control of a Nonlinear Wing Section using Fly-by-Feel Sensing," AIAA Atmospheric Flight Mechanics Conference, 2015.

[4]Kopsaftopoulos, F., Nardari, R., Li, Y.-H., Wang, P., and Chang, F.-K., "Stochastic global identification of a bio-inspired self-sensing composite UAV wing via wind tunnel experiments," Proceedings of the SPIE 9805, Health Monitoring of Structural and Biological Systems 2016, 98051V, April 2016.

[5]Ihn, J. and Chang, F.-K., "Pitch-catch active sensing methods in structural health monitoring for aircraft structures," *Structural Health Monitoring*, Vol. 7, (1), 2008, pp. 5–19.

[6]Zhuang, Y., Kopsaftopoulos, F., and Chang, F.-K., "Bondline integrity monitoring of adhesively bonded structures via an electromechanical impedance based approach," Structural Health Monitoring 2015: System Reliability for Verification and Implementation – Proceedings of the 10th International Workshop on Structural Health Monitoring (IWSHM 2015), edited by F.-K. Chang and F. Kopsaftopoulos, Vol. 2, 2015.

[7]Kopsaftopoulos, F., Nardari, R., Li, Y.-H., and Chang, F.-K., "A stochastic global identification framework for aerospace structures operating under varying flight states," *Mechanical Systems and Signal Processing*, Vol. 98, 2018, pp. 425–447.
doi: 10.1016/j.ymssp.2017.05.001

[8]Kopsaftopoulos, F., Nardari, R., Li, Y.-H., and Chang, F.-K., "Data-driven state awareness for fly-by-feel aerial vehicles: Experimental assessment of a non-parametric probabilistic stall detection approach," Structural Health Monitoring 2017: Real-Time Material State Awareness and Data-Driven Safety Assurance – Proceedings of the 11th International Workshop on Structural Health Monitoring (IWSHM 2017), edited by F.-K. Chang and F. Kopsaftopoulos, September 2017.

[9]Lanzara, G., Salowitz, N., Guo, Z., and Chang, F.-K., "A spider-web-like highly expandable sensor network for multifunctional materials," *Advanced Materials*, Vol. 22, (41), 2010, pp. 4643–4648.

[10]Dowell, E., Edwards, J., and Strganac, T., "Nonlinear aeroelasticity," *Journal of aircraft*, Vol. 40, (5), 2003, pp. 857–874.

[11]Livne, E., "Future of airplane aeroelasticity," *Journal of Aircraft*, Vol. 40, (6), 2003, pp. 1066–1092.

[12]Georgiou, G., Manan, A., and Cooper, J., "Modeling composite wing aeroelastic behavior with uncertain damage severity and material properties," *Mechanical Systems and Signal Processing*, Vol. 32, 2012, pp. 32–43.

[13]Jones, J. and Cesnik, C., "Nonlinear Aeroelastic Analysis of the X-56A Multi-Utility Aeroelastic Demonstrator," Proceedings of 5th Dynamics Specialists Conference, January 2016.

[14]Kitson, R. C. and Cesnik, C. E., "Aeroelastic Modeling and Simulation of High-Speed Flexible Vehicles," 15th Dynamics Specialists Conference, 2016.

[15]Kopsaftopoulos, F. P. and Fassois, S. D., "Vector-dependent Functionally Pooled ARX models for the identification of systems under multiple operating conditions," Proceedings of the 16th IFAC Symposium on System Identification, (SYSID), July 2012.

[16]Kopsaftopoulos, F. P., *Advanced Functional and Sequential Statistical Time Series Methods for Damage Diagnosis in Mechanical Structures*, Ph.D. thesis, Department of Mechanical Engineering & Aeronautics, University of Patras, Patras, Greece, January 2012.

[17]Kopsaftopoulos, F. P. and Fassois, S. D., "A Functional Model Based Statistical Time Series Method for Vibration Based Damage Detection, Localization, and Magnitude Estimation," *Mechanical Systems and Signal Processing*, Vol. 39, 2013, pp. 143–161.
doi: 10.1016/j.ymssp.2012.08.023

[18]Salowitz, N., Guo, Z., Li, Y.-H., Kim, K., Lanzara, G., and Chang, F.-K., "Bio-inspired stretchable network-based intelligent composites," *Journal of Composite Materials*, Vol. 47, (1), 2013, pp. 97–106.

[19]Guo, Z., Kim, K., Lanzara, G., Salowitz, N., Peumans, P., and F.-K.Chang, "Bio-inspired smart skin based on expandable network," Proceedings of the 8th International Workshop on Structural Health Monitoring 2011 - Condition Based Maintenance and Intelligent Structures, edited by F.-K. Chang, September 2011.

[20]Salowitz, N., Guo, Z., Kim, S.-J., Li, Y.-H., Lanzara, G., and Chang, F.-K., "Screen-printed piezoceramic actuators/sensors microfabricated on organic films and stretchable networks," Proceedings of the 9th International Workshop on Structural Health Monitoring 2013, edited by F.-K. Chang, September 2013.

[21]Guo, Z., *Robust Design And Fabrication Of Highly Stretchable Sensor Networks For The Creation Of Intelligent Materials*, Ph.D. thesis, Department of Aeronautics and Astronautics, Stanford University, 2014.

[22]Salowitz, N., Guo, Z., Roy, S., Nardari, R., Li, Y.-H., Kim, S., Kopsaftopoulos, F., and Chang, F.-K., "A vision on stretchable bio-inspired networks for intelligent structures," Proceedings of the 9th International Workshop on Structural Health Monitoring (IWSHM 2013), edited by F.-K. Chang, September 2013.

[23]Sakellariou, J. S. and Fassois, S. D., "Functionally Pooled models for the global identification of stochastic systems under different pseudo-static operating conditions," *Mechanical Systems and Signal Processing*, Vol. 72-73, 2016, pp. 785807. doi: 10.1016/j.ymssp.2015.10.018

[24]Ljung, L., *System Identification: Theory for the User*, Prentice–Hall, second edition, 1999.

[25]Söderström, T. and Stoica, P., *System Identification*, Prentice–Hall, 1989.

[26]Fassois, S. D., "Parametric identification of vibrating structures," *Encyclopedia of Vibration*, edited by S. Braun, D. Ewins, and S. Rao, Academic Press, 2001, pp. 673–685.

[27]Bernstein, D., *Matrix Mathematics*, Princeton University Press, 2005.

[28]Janapati, V., Kopsaftopoulos, F., Li, F., Lee, S., and Chang, F.-K., "Damage detection sensitivity characterization of acousto-ultrasound-based structural health monitoring techniques," *Structural Health Monitoring*, Vol. 15, (2), 2016, pp. 143–161.

[29]Yadav, S., Chung, H., Kopsaftopoulos, F., and Chang, F.-K., "Damage quantification of active-sensing acousto-ultrasound-based SHM via a multi-path unit-cell approach," Structural Health Monitoring 2017: Real-Time Material State Awareness and Data-Driven Safety Assurance – Proceedings of the 11th International Workshop on Structural Health Monitoring (IWSHM 2017), edited by F.-K. Chang and F. Kopsaftopoulos, September 2017.

AUTHOR INDEX

Vertical Flight Society
2701 Prosperity Ave, Suite 210
Fairfax, VA 22031
USA

ISBN 978-1-5108-7001-7

9 781510 870017

CURRAN ASSOCIATES INC.
proceedings.com